Directions in Physics

Professor Dirac at the Department of Physics, University of Canterbury, Christchurch, New Zealand.

Directions in Physics

LECTURES DELIVERED DURING A VISIT TO
AUSTRALIA AND NEW ZEALAND
August/September 1975

P. A. M. DIRAC, F.R.S., O.M.

Professor of Physics
Florida State University
Tallahassee

Edited by

Professor H. Hora and J. R. Shepanski

Department of Theoretical Physics
University of New South Wales
Kensington, Sydney, Australia

with a Foreword by

Professor Sir Mark Oliphant, F.R.S., K.B.E.
Governor of South Australia

A WILEY-INTERSCIENCE PUBLICATION

JOHN WILEY & SONS, New York · London · Sydney · Toronto

Library of Congress Cataloging in Publication Data:

Dirac, Paul Adrien Maurice, 1902-
    Directions in physics.

    "A Wiley-Interscience publication."
    Includes index.
    1. Physics—Addresses, essays, lectures. I. Title.

QC71.D55   530   77-24892
ISBN 0-471-02997-1

Printed in the United States of America

10 9 8 7 6 5 4 3 2

Foreword

Only rarely can the great figures in the physical sciences be persuaded to visit the antipodes. Hence it was remarkable that Professor H. Hora and Dr. J. L. Hughes were able, during a conference in Miami, to make tentative arrangements for the one surviving father of modern quantum theory to travel to Australia and New Zealand and to deliver there the outstanding lectures reproduced in this volume. Professor Dirac lectured in Christchurch, New Zealand and in Sydney, Adelaide and Canberra, Australia.

Those of my generation who were privileged to know Rutherford, Bohr, Heisenberg, Pauli, Chadwick, Schroedinger, Einstein and others, and hear them speak of their latest investigations, did not realize fully at the time that we were in the midst of a revolution in knowledge of the physical world. The contributions of P. A. M. Dirac to that revolution were outstanding, but when he lectured in the Cavendish on the positively charged antiparticle to the electron, we were disappointed that it did not have the mass of the proton! However, it was not long before his ideas were vindicated by the discovery of the positron, and we could then speak with awe of Dirac's uncanny intuition.

It is doubtful whether such a period of concentrated discovery will ever occur again in physics. This is the significance of these lectures—to hear from the creator himself of the way in which the new concepts were generated, and what their consequences are. Some, like the possible variation of the gravitational constant with time or the existence of magnetic monopoles, remain speculative, but the very uncertainty adds to their stimulus for further work.

We are indebted to Dr. Hughes for arranging the visit by Professor Dirac and for his indefatigable pursuit of the funds which made it possible. We thank the institutions and the individuals in Australia and New Zealand who contributed so generously, especially the Universities of New South Wales, Adelaide, Flinders, Canterbury, and the Australian National University. The support of the Australian Institute of Physics and of Quentron Pty. Ltd. is acknowledged with gratitude.

J. R. Shepanski, Lecturer in the Department of Theoretical Physics, University of New South Wales, painstakingly wrote up and edited the lectures from tape recordings, and these were then submitted to Professor Dirac for further editing and approval for publication.

MARK OLIPHANT

v

Professor and Mrs. Dirac on an excursion to Shark Island in Sydney Harbour, organized by the Department of Theoretical Physics, University of New South Wales, Kensington, Sydney, Australia.

Same scene as above. Persons from right to left following Mrs. Dirac: Mrs. Mabbs George, Mrs. Irene Kelly, and Mr. V. F. Lawrence, a PhD student in theoretical physics and a well-known tennis player.

Professor Dirac at his colloquium in the School of Physics, University of New South Wales, on "Cosmology and the Gravitational Constant."

Editorial Remarks

It is difficult to think of a person more qualified or better suited than Professor Dirac to describe both the triumphs and problems of modern physics. A foremost pioneer in the development of quantum mechanics, he still remains very active in this and other frontier fields of physics. In addition, he has the uncanny facility of presenting very complex problems in a readily understandable way, so that a large group of people, comprising not only scientists, engineers, or teachers but also educated laymen, can gain valuable insight from his exposition.

Professor Dirac gave a number of lectures on selected topics during his visit to Australia and New Zealand. Some of the topics were covered in similar talks given in different places. It was rather difficult, therefore, to make the selection for this book.

It was decided, for example, to choose the Sydney version of the lecture on "The Development of Quantum Mechanics" rather than that given in Adelaide (transcripts of the latter are available, together with comments from Professor C. A. Hurst, and copies may be obtained from the Department of Mathematical Physics, University of Adelaide). The Sydney lecture was preceded by the remarks of Professor H. J. Goldsmid, who welcomed Professor Dirac on behalf of the Australian Institute of Physics. These remarks are reproduced together with the lecture. It is characteristic of the acceptance accorded Professor Dirac that an audience of over 500 was present at that lecture, even though all three Sydney universities were then on vacation.

The lectures on "Quantum Electrodynamics" and "Magnetic Monopoles" come from videotape recordings of the presentations at the University of Canterbury, Christchurch, New Zealand. The concept of magnetic monopoles is another of the ingenious ideas of Dirac. This subject was very topical at the time of his visit because of worldwide interest in some experiments indicating the possible existence of such monopoles.

The lecture on "A Positive Energy Relativistic Wave Equation" is of a rather specialized character and is, perhaps, more suitable for experts. This topic was, at the time, the subject of a special discussion at the Department of Theoretical Physics at the University of New South Wales.

For the last lecture, "Cosmology and the Gravitational Constant," we have decided also to include the introductory remarks of Professor E. P. George, Head of the School of Physics, The University of New South Wales, as well as some extracts of the lively discussion which followed.

All of Dirac's audiences felt profound gratitude to him for describing his unique view of the directions in physics. It is hoped that this view may also be transmitted to the readers of these lectures.

THE EDITORS

Contents

Directions in Physics

1 — The Development of Quantum Mechanics

Introductory Remarks by
Professor H. J. Goldsmid

Public lecture organized by the Australian Institute of Physics at the School of Physics, University of New South Wales, Kensington, Sydney, Australia, August 25, 1975. Introduction by Professor H. J. Goldsmid.

As Chairman of the New South Wales Branch of the Australian Institute of Physics, I have the honor and pleasure to welcome you to an outstanding occasion in the history of physics in Australia. For, our lecture today, which has been sponsored by both The University of New South Wales and the Australian Institute of Physics, is to be given by that very eminent physicist, Professor Paul Dirac.

It has been rightly said—and here I am quoting from others who are more competent to give an opinion on the matter than I am—it has been said that: "posterity will rate Paul Dirac as one of the greatest physicists of all time." His reputation was already established by a paper on quantum mechanics, which he wrote at the age of 23, and he became professor of mathematics at Cambridge University when he was only 30. A year later, he was awarded the Nobel Prize. He was already, at that time, a Fellow of the Royal Society of 3 years' standing, and honors have been bestowed on him continuously throughout the succeeding years.

Many of his theoretical predictions have long ago become well-established facts. Thus the positron was observed experimentally 2 years after he forecast its existence. Other predictions, such as the continuous variation of the constant of gravitation, are only now receiving experimental support. Professor Dirac's theories, indeed, form a very real part of physics today.

But you have not come to listen to me. So let me without further ado welcome Professor Dirac and invite him to speak to us on "The Development of Quantum Mechanics," so much of which has, in fact, been due to his own efforts.

Professor Dirac!

PROFESSOR P. A. M. DIRAC

I am very happy to be here, in Sydney, and to have this chance of talking to you. I would like to talk about the development of quantum mechanics; I have lived through a good deal of this development, although it did actually start before my time. Quantum mechanics is a development of the classical mechanics of Newton. Newton set up his laws of mechanics, which have dominated the whole of mechanical theory since his time and which are found to agree very well with large-scale observations, subject to some modifications needed by the Einstein theory of relativity. They work very well, so long as they are

2

applied to large bodies. When one applies them to very small things, like we have in the atomic world, they just fail.

The mechanics of Newton we call classical mechanics—it is our starting point. We can apply this classical mechanics to the motion of electric charges, charged particles, by making use of the theory of Maxwell. The result that we get we see will not apply to atoms.

The picture of an atom which has been established is that there is a central nucleus, containing a positive charge, with one or more electrons moving around it. According to the mechanics of Newton and Maxwell, as applied to electronic charges, these electrons would gradually lose energy by radiation and would, eventually, collapse into the nucleus. The atom would, therefore, not be a stable thing at all. Well, we know that atoms are stable, so there was a contradiction which, of course, bothered people very much, right at the beginning.

The big solution to this contradiction was obtained by Niels Bohr. He said that we must suppose that the atom can exist in certain "stationary" states and, for these stationary states, it does not emit radiation. That means that one must assume a departure from the standard equations of the Newtonian mechanics, applied to electric charges; the forces that are associated with the emission of radiation have to be neglected. But these forces are small forces: they are not important in a first approximation. The important forces are the Coulomb forces between the electric charges. Well, according to Bohr, we have to make this approximation in the equations governing the motion of the electrons in an atom, and then we have to assume that the atom can exist only in certain states, called stationary states. These states are fixed by some conditions of a type which is entirely unfamiliar to classical mechanics: a new set of conditions called quantum conditions, which involve Planck's constant, (denoted by h) which Planck discovered in his law of blackbody radiation.

Bohr set up this theory of a model for the atom, where it exists in stationary states, subject to these quantum conditions. An atom, according to Bohr, can jump from one stationary state to another, and, when it does jump, it emits (or absorbs) radiation, to preserve conservation of energy. This radiation, which is emitted or absorbed, is a single quantum with a definite frequency connected with the energy.

Now, these ideas are very different from what we are taught by Newtonian mechanics: the idea of stationary states, satisfying special conditions, is something very new. But still, these ideas of Bohr were found to be very successful in accounting for the spectrum of hydrogen and other simple atoms, where there is just one electron that is important. The success was so great that one had to accept the Bohr theory.

I remember what a surprise it was to me when I first learned about the Bohr theory. Previously, the whole world of the atom was completely shrouded in mystery. When I was a student in Bristol, I did not hear anything about this theory of Bohr; it was only when I went to Cambridge as a research student that I was told about it, and it then opened my eyes to a new world and a very surprising world. The surprising thing was that we can apply Newton's laws to the motion of the electrons in an atom, subject to certain conditions. These conditions are first, that we have to neglect the forces on the electrons associated with the emission of radiation and second, that we have to bring in quantum conditions. I still remember very well how strongly I was impressed by this Bohr theory. I believe that the introduction of these ideas by Bohr was the greatest step of all in the development of quantum mechanics. It is really the most unexpected, the most surprising thing that such a radical departure from the laws of Newton should be successful.

Well, people worked on this theory of Bohr, and they found that the successes were very limited. Essentially, one had successes when one was dealing with an atomic system for which there was just one electron playing an important part. If one had two electrons, or more than two, such as occurs in the atom of helium or the more complicated atoms, then one did not know how to apply the quantum conditions to fix the stationary states. People did various calculations, making artificial assumptions, but the work was not successful. That was the situation when I first started research on atomic theory. I was faced with the problem that everyone was working on in those days: "How can one develop the idea of Bohr orbits to apply to these more complicated atoms?"

The great advance was made by Heisenberg in 1925. He made a very bold step. He had the idea that physical theory should concentrate on quantities which are closely related to observed quantities. Now, the things you observe are only very remotely connected with the Bohr orbits. So Heisenberg said that the Bohr orbits are not very important. The things that are observed, or which are connected closely with the observed quantities, are all associated with two Bohr orbits and not with just one Bohr orbit: *two* instead of *one*. Now, what is the effect of that?

Suppose we consider all the quantities of a certain kind associated with two orbits, and we want to write them down. The natural way of writing down a set of quantities, each associated with two elements, is in a form like this:

$$
\begin{pmatrix}
\times & \times & \times & \times & \cdot\ \cdot\ \cdot \\
\times & \times & \times & \times & \cdot\ \cdot\ \cdot \\
\times & \times & \times & \times & \cdot\ \cdot\ \cdot \\
\times & \times & \times & \times & \cdot\ \cdot\ \cdot \\
\cdot & \cdot & \cdot & \cdot & \cdot\ \cdot\ \cdot
\end{pmatrix},
$$

an array of quantities, like this, which one sets up in terms of rows and columns. One has the rows connected with one of the states, the columns connected with the other. Mathematicians call a set of quantities like this a matrix.

Now, Heisenberg assumed that one should deal with such a set of quantities and consider the whole set together as corresponding to one of the dynamical variables of the Newtonian theory. These dynamical variables are, of course, the coordinates of the particles, or the velocities, or momenta. Each of these quantities is to be replaced by a matrix, according to Heisenberg. The underlying idea that led Heisenberg to think of this was that one should construct the theory in terms of observable quantities and that the observable quantities are these matrix elements, each associated with two orbits.

Now, when one deals with matrices, one can add them and multiply them and set up an algebra with them, but one gets an important new feature coming in: that is, if one multiplies two of them together, a and b, and forms ab, the result will usually be different from what one gets if one multiplies b by a. There are definite rules for multiplying matrices which one cannot escape from, and these rules result in ab not being the same as ba. If we are going to count our dynamical variables as matrices, it means that dynamical variables will satisfy an algebra in which ab is not equal to ba, which is called a noncommutative algebra.

When Heisenberg first noticed that his matrices did not satisfy commutative multiplication, he was very disturbed by it. He felt that perhaps the whole theory would break down over that point. (From time immemorial, physicists had been working with dynamical variables for which we always have ordinary algebra; a times b equals b times a. And it was inconceivable to have dynamical variables for which this property fails). Heisenberg was naturally very disturbed by it, but still it was a fundamental point in his theory, and it turned out to be a most important point. Actually, the most important point of the new mechanics of Heisenberg is that the dynamical variables are subject to an algebra where we do not have commutative multiplication.

Well, from the initial idea of Heisenberg, one could make a fairly rapid development, and I was able to join in it. I was just a research student at that time. I was lucky enough to be born at the right time to make it possible for that to be so.

We had the problem, you see, of changing the equations of Newton so as to fit in with an algebra where ab is not equal to ba. Now, at first sight it might seem that that would be rather difficult. But the problem was made very much easier by some work of Hamilton which was done a hundred years previously. Hamilton had been studying the equations of Newton, and he found another way of writing them. There was already one general way of writing the equations, due to Lagrange—Hamilton found another way of writing them, called the Hamiltonian form of the equations. In setting up this form of the equations, Hamilton was influenced only by conditions of mathematical beauty. He might have said: "It is very nice to write the equations in this way, but there is no real necessity to write them in this way. You could, if you liked, continue to use the equations in the form that they were originally given by Newton."

But Hamilton seemed to have some remarkable insight into what was important—one of the most remarkable insights, I suppose, which a mathematician has ever had. He found a form of writing the equations of mechanics whose importance would be realized only after a hundred years, that is to say, long after his death.

The importance of the Hamiltonian form of the equations is that one can very easily generalize them to bring in the noncommutation. One can write the Hamiltonian equations in terms of a certain bracket expression called the Poisson bracket, which is usually written like this:

$$[a, b].$$

(I will not write down the definition of this bracket expression. I will just say that it does turn up in the Hamiltonian form of the equations and is of fundamental importance.) And it turns out that this bracket expression, the Poisson bracket, corresponds—is very closely analogous to—a times b minus b times a, divided by the root of minus one times a certain universal constant, crossed h, which is this h of Planck, divided by 2π

$$\frac{ab - ba}{\sqrt{-1}\,\hbar}.$$ (1)

By replacing the Poisson bracket of the Hamiltonian form of the equations by the commutator, $ab - ba$, in accordance with this formula, we can immediately pass from all the equations of classical mechanics (after we have put them into Hamiltonian form) to some new equations where we have noncommutative multiplication and which we can apply to the Heisenberg form of quantum mechanics.

It was then an interesting game people could play to take the various models of dynamical systems, which we were used to in the Newtonian theory, and transform them into the new mechanics of Heisenberg, using this general formula here:

$$[a, b] \to \frac{ab - ba}{\sqrt{-1}\,\hbar}.$$ (2)

It was a good description to say that it was a game, a very interesting game one could play. Whenever one solved one of the little problems, one could write a paper about it. It was very easy in those days for any second-rate physicist to do first-rate work. There has not been such a glorious time since then. It is very difficult now for a first-rate physicist to do second-rate work. But then we had the comparatively easy possibility of passing from the Hamiltonian form of Newtonian mechanics to the new mechanics of Heisenberg, resulting in our having equations belonging to the new quantum mechanics.

We had then equations involving these noncommuting quantities, but to begin with, we had no general interpretation for the equations. That was really a remarkable situation to have in a physical theory. (In any physical theory one usually knows just what one's equations mean before one sets them up. But here, we had a different situation: we had the equations before we knew how to apply them.)

The interpretation was just worked out gradually. One made guesses in simple examples. One could interpret the diagonal matrix corresponding to the total energy as having diagonal elements giving the energies of the states in the quantum theory. And then one gradually worked toward more general methods of interpretation.

I might mention that we had the general equation of motion for any dynamical variable. According to Hamilton, any dynamical variable, u, would vary with the time according to:

$$\frac{du}{dt} = [u, H], \tag{3}$$

where H is the total energy in the Hamiltonian theory. We see that it corresponds to this quantum equation:

$$\frac{du}{dt} = \frac{uH - Hu}{i\hbar}. \tag{4}$$

Here we have the general equation of motion for a dynamical variable in the new mechanics of Heisenberg.

Well, we had to face the problem of getting a general interpretation for these new equations, and this problem was very much assisted by some work of Schrödinger. Schrödinger was working independently of Heisenberg and set up a theory of his own, an alternative form of quantum mechanics, which at first sight appeared to be completely different from the Heisenberg theory. Schrödinger came a little bit later than Heisenberg. After a few months, it was established that the Schrödinger theory and the Heisenberg theory were really equivalent, although they looked so very different in their starting points.

Schrödinger's theory was based on some previous work of de Broglie, who showed how one could introduce waves connected with particles. De Broglie had wave functions which are usually written ("psi"):

$$\psi.$$

For a single particle, ψ would be a function of the three coordinates of the particle, let us call them x_1, x_2, x_3, and the time:

$$\psi(x_1, x_2, x_3, t).$$

De Broglie set up an equation to govern the waves described by ψ, and this wave equation is such that if one takes plane waves moving in a definite direction, with a definite frequency, they correspond to a particle with a definite momentum and definite energy. The correspondence is a relativistic correspondence and is a very neat one mathematically. De Broglie was led to this idea of connecting waves with particles, by mathematical beauty. De Broglie's theory applied only to particles not acted on by any forces. But Schrödinger was able to generalize the theory so as to apply it to an electron moving through an electromagnetic field acted on by electric and magnetic forces.

In Schrödinger's theory, one had operators operating on ψ ("minus ih times partial differentiation with respect to x_r")

$$-i\hbar \frac{\partial}{\partial x_r} = p_r, \tag{5}$$

with r taking the values 1, 2, and 3, and each of these operators corresponds to one of the momentum operators. If one is dealing with such operators, and also with the xs, then one is dealing with noncommuting quantities that are similar to the noncommuting quantities of the Heisenberg theory. And on that basis, one is able to establish the connection between the Schrödinger theory involving these wave functions, with the operators operating on it, and the Heisenberg theory.

Now, the ψ itself did not occur in the original, Heisenberg form of the theory but it got introduced into quantum mechanics through this work of Schrödinger. And then one found that the wave function ψ corresponds to one of the states, perhaps one of the stationary states of the Bohr theory. The operators, which change one wave function to another, are thus connected with two states. And so, the equivalence of the Schrödinger theory and the Heisenberg theory was established.

Now, the general method of interpreting this new mechanics, which was obtained 2 or 3 years after the equations, involved taking this wave function ψ and forming the square of its modulus:

$$|\psi|^2$$

and then assuming this to be the probability of the particle being at a given position at a certain time.

Now, I have introduced here the word probability. That means that the general interpretation of quantum mechanics involves probabilities. This general interpretation enables us to calculate the probability of a certain event, in this case, the probability of the electron being at a certain point at a certain time. According to the Newtonian mechanics, classical mechanics, we do not just calculate probabilities, we calculate precisely which events occur. The new mechanics, quantum mechanics, does not have this determinacy of the Newtonian mechanics. This lack of determinacy is really a great stumbling block toward understanding the new mechanics. It is something that is very hard to accept.

Of course, experimental results with atoms do involve probabilities, and we can calculate probabilities, in accordance with the new mechanics, and compare them with the experimental results. And one then finds that one gets agreement between the calculations and the observations. From that point of view the probabilities are all that one really needs. But still, one does not feel satisfied with a theory which gives only probabilities. This has led to a very great controversy.

Some physicists, led by Einstein, have supposed that, basically, physics should be deterministic and should not merely provide probabilities. Now, Bohr accepted the probability interpretation. He was able to fit it into his philosophy. And that led to a very big controversy between the Bohr school and the Einstein school, a controversy which lasted throughout the life of Einstein. Well, they were both most eminent physicists. The question is, "Who was right of those two?"

It seems that according to the standard accepted ideas of atomic theory, Bohr was right. This interpretation in terms of probabilities, based on the Schrödinger wave function, is the best that one can do. People have made many attempts to try to improve on it, to try to get more information than merely those probabilities. But such attempts have been failures! According to the present quantum mechanics, the probability interpretation, the interpretation which was championed by Bohr, is the correct one. But still, Einstein did have a point. He believed that, as he put it, the good God does not play with dice. He believed that basically physics should be of a deterministic character.

And, I think it might turn out that ultimately Einstein will prove to be right, because the present form of quantum mechanics should not be considered as the final form. There are great difficulties, which I shall mention later, in connection with the present quantum mechanics. It is the best that one can do up till now. But, one should not suppose that it will survive indefinitely into the future. And I think that it is quite likely that at some future time we may get an improved quantum mechanics in which there will be a return to determinism and which will, therefore, justify the Einstein point of view. But such a return to determinism could only be made at the expense of giving up some other basic idea which we now assume without question. We would have to pay for it in some way which we cannot yet guess at, if we are to re-introduce determinism.

Well, those are the ideas governing the fundamental equations of the new mechanics and their interpretation. And I would like now to discuss one particular problem which concerned me very much, namely, how to fit in these equations with the mechanics of the Einstein theory. The Newtonian equations, which I mentioned to begin with, are valid only for particles that do not move very fast, whose speeds are not large, not comparable with the velocity of light. As soon as one gets very rapidly moving particles, one has to go over to a new mechanics, the mechanics of Einstein's special relativity. But this new mechanics is still within the framework of the Newtonian theory and has equations that can be put into the Hamiltonian form. But certain special problems arose, and dealing with those problems led eventually to the concept of antimatter. I would like to discuss the main features of this development.

I must write down a few equations. If you consider the energy of a particle, according to the Newtonian theory, you have

$$E = \tfrac{1}{2}mv^2 = \frac{1}{2m}\,p^2,$$

where p is the particle momentum. According to Einstein, when v is large, when it becomes comparable with the velocity of light, we must replace this formula by

$$E = c\sqrt{m^2c^2 + p^2}. \tag{6}$$

Now, this formula of Einstein looks very different from the formula of Newton. The difference, in the first place, comes from the fact that if the particle is not moving at all, the energy is zero according to Newton; but it is not zero according to Einstein, it is mc^2.

Well, the Einstein theory gives the particle an extra energy, independent of its velocity, called the rest energy, an energy which is supposed to be locked up inside the particle. For small values of the momentum p, the Einstein formula is

$$E = mc^2 + \frac{1}{2m}\,p^2 + \cdots$$

and then we have some more terms here involving higher powers of p. Thus the extra energy, in the case of the particle not moving very fast, is in agreement with the Newtonian theory.

Now there is one other difference between the Einstein formula for the energy and the Newtonian formula, namely, there is a square root in the energy expression (6). Now, whenever you see a square root, you know mathematically that you can have a plus or a minus sign in front of it. So this energy, according to the Einstein formula, can have negative values as well as positive values. One usually writes energies as horizontal lines denoting various levels of energy. Then, according to this formula of Einstein, the energy level can be mc^2 or anything greater than mc^2, going up to infinity. We have a whole lot of energy levels like those in Figure 1. We have another lot of energy levels, starting with minus mc^2, and going down to minus infinity, as seen in Figure 2. These are the energy levels allowed by the Einstein formula shown in Figure 3.

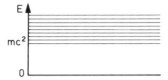

Figure 1. Positive energy levels according to Einstein's formula.

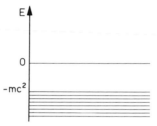

Figure 2. Negative energy levels.

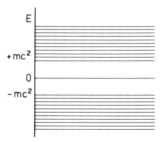

Figure 3. All allowed energy levels from the Einstein formula.

In practice, one always sees only positive energies for particles. So this formula of Einstein, Equation (6), allows values for the energy which are not observed in practice. Now, that did not bother people to begin with. One just said: "let us ignore these negative energies and work entirely with positive energies." It was permissible to do that because, if we start off our particle in a state of positive energy, the energy always remains positive, and these negative energies do not play any role at all in the theory.

Now, Equation (6) applies for a particle in the absence of any external field. We can generalize it to a charged particle, such as an electron, in the presence of an electromagnetic field, and the situation is not very much changed. We can work out the energy levels under those more general conditions, and we have Lorentz's equations, classical equations, which tell us how the particle moves in accordance with the Einstein mechanics when it is acted on by electric and magnetic fields. People worked with these Lorentz equations, applying them to an electron with a positive energy.

Now, one might apply them to an electron with a negative energy. So far as I know, no one ever did that calculation: people were just not interested in the negative energies at all at that time. If one does the calculation and proceeds to examine how the electron would move, if one starts it off in a state of negative energy, using the Lorentz equations, one finds that it always remains in a state of negative energy and it moves as though it had a positive energy and a positive charge (the ordinary electron has a negative charge). Thus it seems as though both the energy and the charge are reversed for these negative-energy states. Well, that is the situation according to classical mechanics with the modifications of the Einstein theory which are needed for large velocities.

When we go over to quantum mechanics the situation is different, because with quantum mechanics we have dynamical variables able to make a jump from one value to another. And, if the energy is started off positive, it does not have to remain positive in the quantum theory; it can jump to a state of negative energy. We could close our eyes to the negative energy states so long as we were dealing with the classical theory. We can no longer do that in the quantum theory.

Well, these negative energies occur in a very fundamental way. But still, people were not bothered by them very much, because they had more profound problems which they were tackling. These problems of understanding the basic ideas of the mechanics and their interpretation were what dominated physicists at that time.

However, one still had to face up to the question of getting a relativistic quantum mechanics. If one worked with the quantum mechanics from the point of view of the waves of de Broglie or Schrödinger, one had a wave function, ψ, and one could set up a relativistic wave equation

$$\left(\frac{1}{c^2}\frac{\partial^2}{\partial t^2} - \frac{\partial^2}{\partial x_1{}^2} - \frac{\partial^2}{\partial x_2{}^2} - \frac{\partial^2}{\partial x_3{}^2} + \frac{m^2c^2}{\hbar^2}\right)\psi = 0. \tag{7}$$

This is the equation that ψ satisfies corresponding to a free particle according to de Broglie, and there is an extension of it showing how the wave function varies with the time when we have an electric or a magnetic field present. I do not want to write down this more complicated equation. I shall just say that it is Equation (7) with some extra terms coming in involving the field quantities.

Now, people had this wave equation for a single particle, for a single electron, in the presence of an electric or a magnetic field, and they were able to set up a relativistic interpretation for the wave function. But that relativistic interpretation was not consistent with the interpretation we have for the general Schrödinger theory. The reason for that failure is that in Equation (7) we have $\partial^2/\partial t^2$, the square of the operation time differentiation, whereas the general Schrödinger theory involves this equation:

$$i\hbar\frac{\partial\psi}{\partial t} = H\psi, \tag{8}$$

which is linear in $\partial/\partial t$.

The result is that, working with Equation (7) and setting up an expression for the probability that is of a relativistic form, we would find that the probability is not always positive. With Equation (8) and with the probability given by $|\psi|^2$, the probability is always positive, as it has to be physically.

There was thus a real difficulty in making the quantum mechanics agree with relativity. That difficulty bothered me very much at the time, but it did not seem to bother other physicists, for some reason which I am not very clear about.

I think that I was so strongly impressed by the beauty and the power of the formalism, based on the equation of motion of Heisenberg (Equation 4) and the corresponding equation of Schrödinger (Equation 8), that I felt one *had* to keep to this formalism and that it would not do to pass over to a different kind of equation where we had $\partial^2/\partial t^2$ instead of $\partial/\partial t$. I remember in particular an incident in the Solvay conference in 1927. During the interval before one of the lectures, Bohr came up to me and asked me: "What are you working on now?" I tried to explain to him that I was working on the problem of trying to find a satisfactory relativistic quantum theory of the electron. And then Bohr answered that that problem has already been solved by Klein (Klein's solution involved Equation 7). I tried to explain to Bohr that I was not satisfied with the solution of Klein, and I wanted to give him reasons, but I was not able to do so because the lecture started just then and our discussion was cut short. But it rather opened my eyes to the fact that so many physicists were quite complacent with a theory which involved a radical departure from some of the basic laws of quantum mechanics, and they did not feel the necessity of keeping to these basic laws in the way that I felt.

Well, I was worrying over this point for some months, and then, ultimately, I found a solution. I found another wave equation:

$$\left\{ i\hbar \left(\frac{\partial}{c\partial t} + \alpha_1 \frac{\partial}{\partial x_1} + \alpha_2 \frac{\partial}{\partial x_2} + \alpha_3 \frac{\partial}{\partial x_3} \right) + \alpha_m mc \right\} \psi = 0, \tag{9}$$

where we now have a ψ involving four components instead of just a single ψ like we have in Equation (8), and these alphas are matrices that operate on the four components. This equation can be shown to be a relativistic equation, and it has the single $\partial/\partial t$ instead of the quadratic $\partial^2/\partial t^2$. Thus it is in agreement with the basic laws of the quantum theory. And this I proposed as the equation for the electron.

Now, I found out that this equation gives the electron a spin of a half a quantum and also gives it a magnetic moment, and this spin and magnetic moment are in agreement with observation. That result was really very surprising, because it meant that the simplest possible solution to the problem of getting a relativistic quantum theory for a particle involved a particle with a spin. I had always assumed that the simplest possible solution would refer to a particle with no spin and that when one had obtained this solution for a particle with no spin, one would then have to introduce the spin later on. Here it turned out that the simplest solution involved a spin.

This worked very well for a good many features of the theory, but the negative energy problem remained unresolved. The new theory still allows negative energies, together with positive energies, and as the other difficulties were solved, this negative energy difficulty remained the prime problem.

The solution to the negative energy problem turns out to be possible if one makes use of another property of electrons. Electrons have the property that we cannot have two of them in the same state. This property is a consequence of the laws of quantum mechanics, when we impose suitable symmetry conditions on the wave function, and it was proposed in the first place by Pauli as a device for accounting for the structure of the atoms in the periodic table of the elements. If you cannot have more than one electron in any state, then there are various shells of electrons that can occur in an atom. The shells get filled up, and then any further electrons that are introduced have to be in an outer shell. We go from one shell to another, and in that way we set up the periodic table of the chemical elements.

Now, with quantum mechanics, we cannot exclude transitions from positive energy states to negative energy states, and that means that we cannot exclude the negative energy states from our theory. If we cannot exclude them, we must find a method of physical interpretation for them. One can get a reasonable interpretation by adopting a new picture of the vacuum. Previously, people have thought of the vacuum as a region of space that is completely empty, a region of space that does not contain anything at all. Now we must adopt a new picture. We may say that the vacuum is a region of space where we have the lowest possible energy. Now, to get the lowest energy we must fill up all the states of negative energy. The more electrons we can put into states of negative energy, the lower the total energy becomes, because each electron in a state of negative energy means a reduction in the total energy. Thus we must set up a new picture of the vacuum in which all the negative energy states are occupied and all the positive energy states are unoccupied.

We can get a departure from the vacuum state in two ways: one way is to bring in an electron into a positive energy state; the other way is to have a "hole" in the distribution of negative energy states. That focuses attention on the holes. Well, one can look into the question of how a "hole" will move if there is an electromagnetic field present. And, it moves in roughly the same way as the electron that fills up that "hole" would move. This is in accordance with the quantum modification of the classical Lorentz equation applied to a negative energy particle. And, as I said before, this negative energy particle, according to the classical Lorentz equation, will move as though it had a positive energy and a positive charge. Thus these "holes" move as though they had positive energies and positive charges instead of the usual negative charge of the electron; the "holes" appear as a new kind of particle having a positive charge.

What is the mass of these new particles? Well, when I first thought of this idea, it occurred to me that the mass would have to be the same as that of the electron because of the symmetry. But I did not dare to put forward that idea, because it seemed to me that if this new kind of particle (having the same mass as the electron and an opposite charge) existed, it would certainly have been discovered by the experimenters. At that time the only particles that were known were the negatively charged electron and the positively charged proton, and the other atomic nuclei which were believed to be composite things. So I put forward the idea that these "holes" correspond to positively charged protons, and I left it as an unsolved problem why they should have such a widely different mass from the mass of the electron.

That, of course, was really quite wrong of me; it was just lack of boldness. I should have said in the first place that the "hole" would have to have the same mass as the original electrons. Other people did point that out pretty soon after my paper was published. I think that Weyl was the first to make the very definite statement that mathematical symmetry demanded that these "holes" should be particles with the same mass as the mass of the electron. (Weyl was a mathematician who was only interested in questions of mathematical symmetry and was not bothered by the fact that this kind of particle was never observed by a physicist.) Well, it seems that this was the correct idea; these are new particles which are now called positrons, which have the same mass as the electron and the opposite charge. And the question arises: "Why had experimenters never observed them?" I think the only answer to that question is that they were prejudiced against new particles.

It was assumed that there were only two basic particles in Nature: the electron and the proton. Only two were needed, because there were only two kinds of electricity: one negative, one positive. If one has one particle for the negative charge, one for the positive charge ..., well, that is sufficient; we do not need any other particles. This was the idea that was prevailing at the time.

They had never observed positrons, because they really turned a blind eye to them when they had evidence for them. You see, one can observe the tracks of charged particles in a Wilson chamber, and if there is a magnetic field present, the track is curved. Now, the track would be the same if one had a positively charged particle moving in one direction as if one had a negatively charged particle moving in the opposite direction. One may assume that all the tracks one observes as electrons correspond to negatively charged electrons and that they are all moving in the appropriate direction for the negative charge. It had been noticed by some ex-perimenters that when one had a radioactive source, one had fairly often particles moving into the source. In fact, there are even one or two published photographs of tracks of particles that are moving into the source, in accordance with that interpretation. No one had ever bothered to make statistics. If they had done so, they would have noticed that there are far too many particles moving into the source for this explanation to be possible. Still, that was the situation in those days; people were very reluctant to postulate a new particle.

The situation is quite different nowadays, when people are only too willing to postulate a new particle on the slightest evidence, either theoretical or experimental. Then, it needed a few years for the positron to get established. Blackett had some rather convincing evidence for the existence of positrons, and he told me about it in Cambridge. But he hesitated to publish such a revolutionary idea: he wanted confirmation of his experiments. And with this delay, the discovery was really scooped by Anderson. Anderson had only one photograph of such a track (shown in Figure 4) passing through a lead plate, and the track was more curved on one side of the plate than on the other side. Now the particle would certainly lose energy in passing through this plate—it could not gain energy. It was thus established quitely definitely in which direction the particle was moving. And this direction corresponded to the particle having a positive charge.

Well, this discovery confirmed the idea that electrons have an antiparticle associated with them, the positron. This idea is really quite a general idea in atomic theory. For all these particles, such that we

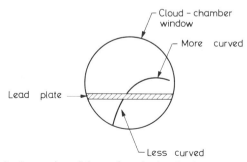

Figure 4. Anderson's observation of the positron in the cloud chamber passing through lead (sketch only).

cannot have more than one in any state—particles which are called fermions—for all such particles we have negative-energy states as well as positive-energy states. The negative-energy states are all filled up in the vacuum and are nearly all filled up in practice. Any "hole" in the negative energy states corresponds to an antiparticle. Now, this applies to protons—there are antiprotons; it applies to neutrons—there are antineutrons; they have been observed.

With the idea of antimatter we have to rather change our views about what is meant by a fundamental particle, or an elementary particle. Particles can be created just out of some other form of energy, such as the energy of electromagnetic waves. We may have electromagnetic waves disturbing the vacuum and pulling up one of the negative energy electrons into a positive energy state, and then we have created an electron and a positron. The electron and the positron get created simultaneously. There is, of course, conservation of electric charge and the energy must come from the outside source. A similar situation exists with the other fermions.

If one can create particles, then the question of which are the fundamental constituents of matter ceases to have a definite meaning. Previously, one could say that one only had to analyze a piece of matter, as far as possible, and get at the ultimate constituents in that way. But, if one can create particles by atomic interactions, then one cannot give a definite definition for an elementary particle. With the modern developments of physics, a very large number of particles have been discovered. Instead of just the two, electron and proton, we now have, may be two hundred or so. Many of them are very unstable, some extremely so, others less so.

If one asks which of these new particles are the elementary ones, one really cannot be given a definite answer. One might be inclined to think that the more stable particles are elementary, the less stable ones are not elementary, but that is a rather artificial separation. For example, we have the proton and the neutron which are very similar in many respects. But the proton is stable, and the neutron is not stable. They are so similar that it would be unreasonable to look on one of them as more elementary than the other. Thus the question of which particles are elementary is one of the unsolved problems which physicists are now concerned with.

Well, I have shown how quantum mechanics developed, and I have discussed in particular this problem of fitting in the theory with the mechanics of Einstein, which is necessary for large velocities. And this fitting in leads to the idea of antimatter. The problems of quantum theory are, however, not solved by this work. There are many problems remaining which are centered on the question of getting a precise theory for the interaction between a charged particle and the electromagnetic field.

You may work with a model for a charged particle in which the charge is considered as concentrated in a point. If you work with such a model, then you find that the energy associated with the point charge is infinitely great. That is typical of the difficulties which occur when one tries to get a precise theory for the interaction of particles.

Our present quantum theory is very good, provided we do not try to push it too far—we do not try to apply it to particles with very high energies and we do not try to apply it to very small distances. When we do try to push it in these directions, we get equations which do not have sensible solutions. We have our interactions always leading to infinities. This question has bothered physicists for 40 years, and they have not made any very substantial progress.

It is because of these difficulties that I feel that the foundations of quantum mechanics have not yet been correctly established. Working with the present foundations, people have done an awful lot of work in making applications in which they can find rules for discarding the infinities. But these rules, even though they may lead to results in agreement with observation, are artificial rules, and I just cannot accept that the present foundations are correct.

I feel that the situation at the present time with regard to these infinities is very similar to what it was when people were working with the wave equation with $\partial^2/\partial t^2$. People are, I believe, too complacent in accepting a theory which contains basic imperfections, and a true advance will be made only when some fundamental alteration is made, just about as fundamental as passing from Equation (7) to Equation (9).

2 — Quantum Electrodynamics

Lecture presented at the Department of Physics, University of Canterbury, Christchurch, New Zealand, September 15, 1975 and organized by Professor B. G. Wybourne, Acting Head of the Department. Professor Dirac's visit to the University of Canterbury was made possible by the award of an Erskine Prestige Fellowship. It is interesting to note that John Angus Erskine, whose bequest to the University of Canterbury made this award possible, was a student at the university at the same time as Rutherford.

I would like to talk today about some developments of quantum mechanics. In my earlier talk, I gave the basic structure of quantum mechanics. That basic structure involves dynamical variables, which are noncommuting quantities, and we have to be given the commutation relations between them in order to make them definite. When we have these definite dynamical variables, we then need a Hamiltonian, that is to say, a quantity which represents physically the total energy and which is a function of these dynamical variables.

When we have got the Hamiltonian (which I denote here by H), we can proceed to set up the Heisenberg equations of motion. Thus for any dynamical variable u we have:

$$i\hbar \frac{du}{dt} = uH - Hu,$$ (1)

which gives us the Heisenberg scheme of equations. Alternatively, there is the Schrödinger formalism, in which we work with a wave function ψ, satisfying the wave equation:

$$i\hbar \frac{d\psi}{dt} = H\psi.$$ (2)

H is again the same Hamiltonian as in (1), but it is now interpreted as an operator operating on the wave function ψ.

I want to apply this now to a dynamical system containing many similar particles. The wave function of that system will involve dynamical variables of each of these particles. We shall consider first the possibility of the wave function being symmetrical between the particles. Now, the Hamiltonian of the system has to be symmetrical between the particles: it represents their total energy, and there is simply nothing to distinguish one particle from another. It follows, therefore, from (2) that if ψ is symmetrical, then $d\psi/dt$ will be symmetrical, which requires that if ψ is initially symmetrical, it will always remain symmetrical.

When this situation arises it would appear as a law of Nature [for these particles] that only symmetrical wave functions occur.

We do, indeed, have such a law of Nature for certain kinds of particles. Such particles are called

Bosons.

They follow a different statistics from the classical statistics. Their statistics was first worked out by Bose, with the help of Einstein.

One finds that photons, the light quanta, have to satisfy this Bose–

Einstein statistics, because this statistics then leads to Planck's law for the blackbody radiation. Thus we assert that photons are, in fact, bosons.

We have our wave function ψ as a function of dynamical variables of the different particles:

$$\psi(q^a, q^b, q^c, \ldots), \tag{3}$$

with q^a for the first particle, q^b for the second, and so on, where q is a single letter denoting all the commuting variables needed for describing a state of that particle. We specify a point in the domain of the wave function (3) by specifying all these qs.

It is sufficient just to specify the qs that occur in (3) without mentioning their order when the wave function is symmetrical. If it were not symmetrical, then this would not do, because then we would have to distinguish which particles were in which states. When ψ is symmetrical, however, that distinction is no longer needed, and it is sufficient merely to describe which states are occupied and how many bosons are in each of them.

We get in this way a possibility of transforming ψ to new variables:

$$\psi(n^1, n^2, n^3, \ldots), \tag{4}$$

where n^1 is the number of qs equal to the first value ("q^1", say), n^2—the number of qs equal to the second value ("q^2"), and so on. We can then use these ns as new dynamical variables. We have in that way a new description, in terms of these new variables n. Each of these ns denotes the number of bosons in a certain state. Each of them is a dynamical variable which has for its eigenvalues 0, 1, 2, 3, and so on. They are just integral eigenvalues. The ns commute with each other: if we specify the number of bosons in one state, this does not interfere with our specifying also the numbers in other states.

Now let us study one of the variables n whose eigenvalues are the integers 0, 1, 2, 3, and so on. We immediately see a connection between that n and the energy of a harmonic oscillator. Indeed, a harmonic oscillator has a set of energy levels whose magnitudes form an arithmetical progression, and we can choose our numerical coefficients so that the differences between the successive energy levels are equal to 1. There is also a zero-point energy of half a quantum for a harmonic oscillator. Let us subtract that energy out, and then we are left simply with the energy levels 0, 1, 2, 3, and so on, corresponding exactly to the eigenvalues of one of our ns. This means that we can picture each of these ns in terms of a harmonic oscillator.

The most convenient way of describing a harmonic oscillator is by means of the

Fock Representation.

I merely say here what the basic ideas of that representation are.

Let us here deal with a harmonic oscillator, a different problem from our earlier boson problem. The energy of the oscillator (its Hamiltonian) H is equal to

$$H = \tfrac{1}{2}(p^2 + q^2) - \tfrac{1}{2} \tag{5}$$

where we have subtracted the zero-point energy, and made the simplification of putting $\hbar = 1$, $m = 1$, $\omega = 1$, thus cutting out all these unnecessary numerical coefficients. The basic state of this oscillator has the wave function, which we may call ψ_0:

$$\psi_0 = e^{-q^2/2}. \tag{6}$$

This ψ_0 represents the normal or unexcited state of the oscillator.

Let us now introduce the variable

$$\eta \equiv \frac{1}{\sqrt{2}}(p + iq), \tag{7}$$

a complex variable, p and q being real. It has for its conjugate:

$$\bar{\eta} \equiv \frac{1}{\sqrt{2}}(p - iq). \tag{8}$$

We may work out from (7) and (8) and using the standard quantum condition

$$qp - pq = i, \tag{9}$$

(with $\hbar = 1$) the quantity $\bar{\eta}\eta - \eta\bar{\eta}$. We find

$$\bar{\eta}\eta - \eta\bar{\eta} = \tfrac{1}{2}(p - iq)(p + iq) - \tfrac{1}{2}(p + iq)(p - iq) = \tfrac{1}{2}(-2i)(qp - pq) = 1. \tag{10}$$

This then means that $\bar{\eta}$ is, in fact, equivalent to the operator of differentiating with respect to η: [since $(\partial/\partial\eta)\eta f - \eta(\partial/\partial\eta)f = f$]. Thus

$$\bar{\eta} \equiv \frac{\partial}{\partial\eta}. \tag{11}$$

The Fock representation consists in working with η and $\bar{\eta}$.

Let us now take ψ_0 [from (6)] and operate on it with $\bar{\eta}$ [using (8)]. Since we have

$$p = -i\frac{\partial}{\partial q},$$

we obtain

$$\bar{\eta}\psi_0 = \frac{1}{\sqrt{2}}(p - iq)\psi_0 = \frac{1}{\sqrt{2}}\left(-i\frac{\partial}{\partial q} - iq\right)\psi_0 = \frac{-i}{\sqrt{2}}\left(\frac{\partial}{\partial q} + q\right)\psi_0$$

$$= \frac{-i}{\sqrt{2}}\left(\frac{\partial\psi_0}{\partial q} + q\psi_0\right) = 0, \tag{12}$$

so that $\bar{\eta}$ applied to ψ_0 gives 0.

Let us next consider some function of η and $\bar{\eta}$ expressible as a power series in these variables. Then, using the commutation relationship (10) to shift all the $\bar{\eta}$ factors to the right and applying that function to ψ_0, we are left with some function g of the ηs only applied to ψ_0:

$$f(\eta, \bar{\eta})\psi_0 = g(\eta)\psi_0, \tag{13}$$

in view of the result (12). It, therefore, follows that the only independent wave functions that are left are

$$\psi_0, \eta\psi_0, \eta^2\psi_0, \eta^3\psi_0, \ldots \tag{14}$$

and so on.

We also find that the energy H [in (5)] is simply

$$H = \eta\bar{\eta} \tag{15}$$

[indeed: $\eta\bar{\eta} = \frac{1}{2}(p + iq)(p - iq) = \frac{1}{2}(p^2 + q^2) + \frac{i}{2}(qp - pq) = \frac{1}{2}(p^2 + q^2) - \frac{1}{2} = H$]. If we now apply this H to the state $\eta^r\psi_0$, we get, on using the commutation rule (10):

$$H\eta^r\psi_0 = (\eta\bar{\eta})\eta^r\psi_0 = \eta(1 + \eta\bar{\eta})\eta^{r-1}\psi_0 = \eta^r\psi_0 + \eta^2(1 + \eta\bar{\eta})\eta^{r-2}\psi_0$$

$$= \cdots = r\eta^r\psi_0 + \eta^{r+1}\bar{\eta}\psi_0,$$

so, that in view of (12), we have

$$H\eta^r\psi_0 = r\eta^r\psi_0. \tag{16}$$

This means that the state $\eta^r \psi_0$ is that stationary state of the oscillator for which the energy, excluding the zero-point energy of a $\frac{1}{2}$ unit, is just r units. If you like, you could say [looking at (14)] that ψ_0 is the unexcited state, $\eta \psi_0$ is the first excited state, $\eta^2 \psi_0$ the second excited state,..., $\eta^r \psi_0$ the rth excited state ... and there you have all the different excited states of the oscillator represented in a very simple way.

Well, this representation of Fock is most useful in our present connection with the bosons. We now *introduce one oscillator for each boson state*; note: not one oscillator for each boson—that would be quite wrong. Now, there might be n bosons in a boson state, and we make that correspond to the oscillator (corresponding to that boson state) being in its nth excited state. In that way, any state for the assembly of bosons can be connected with a state for the oscillators. The number of bosons in any boson state is equal to the degree of excitation of the corresponding oscillator.

This is rather a remarkable fact, and it forms the basis of the reconciliation of the wave and corpuscular theories of light. If we look on light from the corpuscular point of view, then we have the photons, which are bosons, and they have to be handled according to this general theory of bosons. If we look on light from the wave point of view, the different Fourier components of the waves are harmonic oscillators, and they may be handled by the Fock treatment. We see now a connection between these two treatments. We see that an assembly of bosons and a set of oscillators are just two mathematical descriptions for the same physical reality. The electromagnetic field may be considered either as an assembly of photons or as a set of electromagnetic waves.

We have these variables η and $\bar{\eta}$ occurring in the Fock treatment. There is a simple physical meaning for them. The η operator increases the degree of excitation by one quantum, whereas $\bar{\eta}$ reduces the degree of excitation by one quantum. We can now understand Equation (12): $\bar{\eta} \psi_0 = 0$. For, if we take the unexcited state [i.e., ψ_0] and try to reduce the degree of excitation by one quantum, we get zero. This is immediately evident physically.

We now have a mathematical description of the electromagnetic field in terms of these operators of increasing or decreasing the degree of excitation of a component by one quantum. They can also be described as the operators of emission and absorption of a boson. The ηs are the emission operators, which increase the excitation by one, and the $\bar{\eta}$s are the absorption (or destruction) operators, which reduce the excitation by one:

$$\text{operators} \begin{cases} \eta \rightarrow \text{emission} \\ \bar{\eta} \rightarrow \text{absorption (destruction).} \end{cases} \tag{17}$$

We have one pair of variables, η^a, $\bar{\eta}^a$, for each boson state. What are the commutation relations between them? They consist of the following:

(i) variables referring to different boson states commute with each other:

emission operators: $\quad \eta^a \eta^b - \eta^b \eta^a = 0,$ (18)
all commute

absorption operators: $\quad \bar{\eta}^a \bar{\eta}^b - \bar{\eta}^b \bar{\eta}^a = 0;$ (19)
all commute

(ii) $\bar{\eta}^a \eta^b - \eta^b \bar{\eta}^a$ equals zero when a and b are different and equals one when a and b are the same, so we write:

$$\bar{\eta}^a \eta^b - \eta^b \bar{\eta}^a = \delta^{ab}, \tag{20}$$

where δ^{ab} is the Kronecker delta symbol. When we are dealing with the electromagnetic field, or any assembly of bosons, we need for the description of the system in quantum mechanics these variables η and $\bar{\eta}$, satisfying the above commutation relations.

Now, I have been talking so far about an assembly of bosons. One might, equally well, have an assembly of particles which are all similar and whose overall wave function is antisymmetrical between them (instead of being symmetrical). Take then the ψ in (2) to be such an antisymmetrical wave function. Then, if this ψ is initially antisymmetrical, it will always remain antisymmetrical. Thus we have a new possibility, referring to a new kind of particle. Particles of this kind are called

Fermions.

They are such that two of them cannot be in the same state. If ψ is antisymmetrical [cf form of expression (3)], we must not take two of the qs equal or we get zero. The property of fermions that two of them cannot be in the same state (called the Pauli exclusion principle), applies to electrons and also to several other kinds of the elementary particles in Nature.

We may proceed now to develop a theory of fermions which is analogous to the foregoing theory of bosons, but the fermion variables occurring in the theory will not be related in any way to harmonic oscillators. Even in the case of the fermions we can, however, introduce operators η and $\bar{\eta}$ which refer to the emission and to the absorption of a fermion, just like the above η and $\bar{\eta}$ operators referred to the emission and absorption of a boson.

I do not have time to go into the details of the development of this theory of fermions; I shall just say what the results are. Again, we have each individual η and $\bar{\eta}$ operator associated with a fermion state. These ηs and $\bar{\eta}$s now satisfy different commutation relations from those for the boson operators, namely,

$$\eta^a\eta^b + \eta^b\eta^a = 0 \tag{21}$$

$$\bar{\eta}^a\bar{\eta}^b + \bar{\eta}^b\bar{\eta}^a = 0, \tag{22}$$

and

$$\bar{\eta}^a\eta^b + \eta^b\bar{\eta}^a = \delta^{ab}, \tag{23}$$

which are equations of the same form as (18), (19), and (20), respectively, except for the plus signs now replacing the minus signs there.

This seems to me a very remarkable mathematical fact. We do not know what really lies behind it. For these are two quite different physical situations. Here, in equations (18) to (20), we are dealing with particles such that any number of them could go into any one state. Here, in equations (21) to (23), we are dealing instead with particles such that two of them cannot be in the same state. These are very different physically, yet there is such a close parallel between the algebraic equations in the two cases.

Suppose now that we take equation (21) and put in it $b = a$. We then get

$$(\eta^a)^2 = 0 \quad \text{[not summed over index } a]. \tag{24}$$

This simply means that we cannot have two fermions emitted into the same state. If we try to do this, then we simply get zero for our wave function. Equation (24) does not have anything corresponding to it in the case of bosons [for, if we put $b = a$ in equation (18), we simply get the identity: $(\eta^a)^2 = (\eta^a)^2$].

The number of particles in a state may be represented, in both cases [that of bosons and that of fermions], by the operator:

$$n^a \equiv \eta^a\bar{\eta}^a \quad \text{[with no summation over index } a]. \tag{25}$$

In the boson case [where the η^a, $\bar{\eta}^a$ operators satisfy conditions (18) to (20)], this operator has the eigenvalues 0, 1, 2, 3, and so on. In the fermion case [where the conditions (21) to (23) apply instead], the operator n^a has the eigenvalues zero and one only. Thus the number of fermions in any one state is either zero, when the state is unoccupied, or is equal to one, when the state is occupied. It is then full so that no other [similar] fermion can jump into it.

These, then, are the basic dynamical variables which are used in any quantum theory of fields. There is, however, a technical generalization one has to make. You see, I have been assuming in the foregoing that the different states [whether for bosons or for fermions] are all discrete. This would not, however, hold in practice. One would usually take the different states to be momentum eigenstates for the particle. The index a denoting the state would then be replaced by the three momentum components for the particle, plus the spin value if the particle has spin. When we do that, the Kronecker δ^{ab} symbol, which we had in equations (20) and (23), will have to be replaced by

$$\delta(p_1' - p_1'')\delta(p_2' - p_2'')\delta(p_3' - p_3'') \tag{26}$$

(in absence of spin), where (p_1', p_2', p_3') and (p_1'', p_2'', p_3'') represent the momenta of the two states. This is just a technical generalization, which we always have to make when we pass from discrete quantum states to a continuous range of quantum states. And this answers the first part of our problem, namely, that of deciding what dynamical variables we are to use in describing the quantum theory of the field.

The next question is to find the Hamiltonian. The Hamiltonian is the total energy of the system, and it has to be chosen so as to give correctly the equations of motion of the system. I would like now to illustrate how that is done by taking the case of a field of radiation by itself.

Take the electromagnetic field, and suppose that there are no charges present, so that we just have a field of photons. We may then use, as our dynamical variables, the electric and magnetic fields. We have to take the dynamical variables throughout the whole of space at one particular time. Thus we take the electric vector $\mathscr{E}$ (for example) at any point (x_1, x_2, x_3) in the three-dimensional space and at certain time t:

$$\mathscr{E}_r(x_1, x_2, x_3; t), \qquad r = 1, 2, 3, \tag{27}$$

and similarly, we take the magnetic field vector $\mathscr{H}$:

$$\mathscr{H}_r(x_1, x_2, x_3; t), \qquad r = 1, 2, 3. \tag{28}$$

These, then, are the dynamical variables that are needed to describe the electromagnetic field of pure radiation.

You will notice that we have rather departed from the four-dimensional symmetry which we like to have in a relativistic theory. This is inevitable whenever we go over to a Hamiltonian formalism: we *must* depart from the four-dimensional symmetry and we just cannot help it, because all our dynamical variables are taken at a certain particular time.

The Hamiltonian is equal to the total energy, and we take it to be the same as in the classical theory:

$$H = \frac{1}{8\pi} \int [\mathscr{E}^2(x_1, x_2, x_3) + \mathscr{H}^2(x_1, x_2, x_3)] dx_1\, dx_2\, dx_3. \tag{29}$$

What are the equations of motion for the variables $\mathscr{E}$ and $\mathscr{H}$? Well, we know what they are, from the Maxwell theory. We have

$$\frac{\partial \mathscr{E}}{\partial t} = \text{curl } \mathscr{H} \tag{30}$$

and:

$$\frac{\partial \mathscr{H}}{\partial t} = -\text{curl } \mathscr{E}. \tag{31}$$

We also have the equations:

$$\text{div } \mathscr{E} = 0, \tag{32}$$

$$\text{div } \mathscr{H} = 0, \tag{33}$$

representing constraints which have to be applied to these variables.

The equations (30) and (31) tell us how $\mathscr{E}$ and $\mathscr{H}$ vary with time. They must correspond to the Heisenberg equations of motion in the quantum theory. Thus in the quantum theory we need

$$i\hbar \text{ curl } \mathscr{H} = \mathscr{E}H - H\mathscr{E} \tag{34}$$

and

$$i\hbar \text{ curl } \mathscr{E} = H\mathscr{H} - \mathscr{H}H. \tag{35}$$

We can infer from these Heisenberg equations of motion what the commutation relations are between the quantities $\mathscr{E}$ and $\mathscr{H}$. The commutation relations have to be such that the equations (34) and (35) follow when we take, for the Hamiltonian H, the expression (29). This is quite a definite problem which we can easily work out.

The result is that the various components of $\mathscr{E}$ throughout space all commute with each other and also the various components of $\mathcal{H}$ throughout space all commute with each other. A component of $\mathcal{E}$ does not commute with a component of $\mathcal{H}$ perpendicular to it if the two field points, that of $\mathcal{E}$ field and that of $\mathcal{H}$ field, are very close together. We have in the commutator the derivative of a delta function coming in. A typical relation is

$$[\mathcal{E}_1(x_1'x_2'x_3'), \mathcal{H}_2(x_1''x_2''x_3'')]_- = \delta(x_1' - x_1'')\delta(x_2' - x_2'')\frac{\partial}{\partial x_3'}\delta(x_3' - x_3'').$$

In any case, the commutator of one of the $\mathcal{E}_r$ with one of the $\mathcal{H}_s$ is simply a number.

We are again able to introduce the variables η and $\bar{\eta}$, referring to the emission and absorption of a photon. In that way we get a satisfactory quantum theory for the electromagnetic field of pure radiation.

We can do the same with an assembly of electrons. We can, furthermore, introduce interaction between the electrons and the electromagnetic field. For this we put down, first, the Hamiltonian that represents the energy of the electrons alone plus that which represents the electromagnetic field alone and then bring in the interaction by adding extra terms to the Hamiltonian, so as to lead to the correct equations of motion. The result is that for an assembly of electrons, together with the positrons which go with them, interacting with an electromagnetic field, we can obtain quite definitely the Hamiltonian that gives the total energy of the dynamical system.

The Hamiltonian thus consists of terms which represent the energies of the particles alone plus terms which represent the interaction energy:

$$H_{\text{total}} = H_{\text{particles}} + H_{\text{interaction}}. \tag{36}$$

The usual way of handling such a Hamiltonian is by a perturbation method. One then looks at the interaction part of the Hamiltonian as giving rise to transitions, namely, transitions in which some particles are emitted and some absorbed.

For each of these transitions there is conservation of momentum, but there is not conservation of energy. You may wonder why there is this distinction here between energy and momentum. The difference arises simply because we are talking here about the energy *at one instant of time*. The Hamiltonian H in (36) applies just to one instant of time, while it includes an integration over the whole of three-dimensional space. Now, whenever we talk about only one instant of time, energies cannot be well defined. Thus the energies do not have to be conserved when a transition is made.

There is one further development of this theory that I would like to point out here, as it is very neat and satisfying. In the first place, one obtains the Hamiltonian not actually in terms of the electromagnetic field-vector variables $\mathscr{E}_r$, $\mathscr{H}_r$, which we had in the discussion, but rather in terms of the four-potential components A_μ, $\mu = 0, 1, 2, 3$. Now, the Hamiltonian in (29) represents the energy of a field of pure radiation, with only transverse electromagnetic waves. As long as we are dealing only with transverse waves, we cannot bring in the Coulomb interactions between particles. To bring them in, we have to introduce longitudinal electromagnetic waves and include them in the potentials A_μ.

The longitudinal waves can be eliminated by means of a mathematical transformation. We like to eliminate them, because they are only rather remotely connected with experiment. The result of this elimination is to give some new variables, replacing the old η and $\bar\eta$ operators, for the emission and absorption of electrons. These new variables have quite a simple physical meaning. Each new variable η refers to the emission of an electron *together with* the Coulomb field around it, not just a *bare* electron. Similarly, each $\bar\eta$ annihilates an electron and its Coulomb field. We thus get a new version of the theory, in which the electron is always accompanied by the Coulomb field around it. Whenever an electron is emitted, the Coulomb field around it is simultaneously emitted, forming a kind of *dressing* for the electron. Similarly, when an electron is absorbed, the Coulomb field around it is simultaneously absorbed.

This is, of course, very sensible physically, but it also means a rather big departure from relativistic ideas. For, if you have a moving electron, then the Coulomb field around it is not spherically symmetrical. Yet it is the spherically symmetric Coulomb field that has to be emitted here together with the electron.

Now, when we do make this transformation which results in eliminating the longitudinal electromagnetic waves, we get a new term appearing in the Hamiltonian. This new term is just the Coulomb energy of interaction between all the charged particles:

$$\sum_{(1,2)} \frac{e_1 e_2}{r_{12}} \tag{37}$$

where the summation extends over all pairs of particles existing in the state of the system with which we are concerned. This term appears automatically when we make the transformation of the elimination of the longitudinal waves.

Well, this part of quantum electrodynamics is really quite satisfactory. I have been able to give you here only a brief outline of it, for there is not much time. Still, one could fill in the details and make everything seem correct. When we go on, however, to consider getting *solutions* of our equations, then we run into problems.

If we want to get solutions, we work from the Schrödinger equation (2):

$$ i\hbar \frac{\partial \psi}{\partial t} = H\psi. \tag{38} $$

The natural way of getting solutions is the perturbation method, in which all the interaction terms are treated as the perturbation.

One tries to get a solution on those lines by starting with some definite initial state. The first-order corrections then work out all right, but, when one tries to calculate the second-order corrections, one is led to integrals which turn out to be infinite. Whichever initial state one chooses, one always gets these infinite integrals appearing in the course of the solution.

This situation is really very disturbing, and I think that the proper conclusion to come to is that this Schrödinger equation (38) has no solutions. At any rate, no one has found a solution, although people have studied it for decades, and, I think, the answer is that it just does not have solutions.

Suppose that we take a very simple case: suppose that we start with a state for which there are no electrons, no positrons, and no photons, thus no particles present at all. If we then apply the perturbation method, we find that, starting with no particles, we do get particles created. This is so because our Hamiltonian in equation (38) contains terms which correspond to the simultaneous creation of an electron, a positron, and a photon. These three things are all created together, with conservation of momentum but without conservation of energy. The result is that our initial state does not remain a state with no particles. Particles are created in the first approximation (the first-order perturbation). Then, when one goes to the second approximation, one gets infinities. Thus, one does not get a solution of (38) even if one starts with this very simple case.

Now, one might say that where there are no particles present, that is the vacuum. That cannot be the case here, because the vacuum ought to be a stationary state. The vacuum must be a state with a lot of particles present corresponding to some stationary solution of the Schrödinger equation. But there are no known solutions of this Schrödinger equation—not even a solution which could represent the vacuum.

One might say that this is a very unfortunate situation and that we cannot do anything at all with this theory. The situation is not, however, really as bad as that, because, from the experimental point of view, we do not want to do calculations about the vacuum. There is nothing the experimenter can do which would give us any information to compare with our calculations about the vacuum. Experimenters are only concerned with departures from the vacuum.

We may depart from the vacuum by taking the ψ for the vacuum state (ψ^{vacuum}, say) and applying to it an operator of emission of an electron:

$$\eta^{(e)}\psi^{\text{vacuum}}. \tag{39}$$

We do not know (as was just pointed out) what ψ^{vacuum} is, but we can apply the Heisenberg equations of motion to this operator $\eta^{(e)}$ of emission of an electron. That would then tell us how this thing (39) varies with the time.

The difficulty with the infinities is not quite so bad under these conditions but, even so, in the second approximation, when we try to solve the Heisenberg equations, we get an infinity coming in. This infinity can be interpreted as an extra self-energy for the electron, self-energy which happens to be infinitely great. That leads to the idea of (mass) *renormalization*:

Renormalization:

We could say that the mass of the electron, which one puts into the equations at the beginning, is not the same as the observed mass. Then, when we bring in the interaction of the electron with the electromagnetic field, that interaction changes the mass and gives it a different value from that of the original mass parameter in the equations of motion.

Now, that is quite a reasonable physical idea if the change in the mass is small or, even if it is not small, if it is finite. It is, however, very hard to attach any sense to it when the change in mass is infinitely great. It is true, though, that the infinity which one gets from trying to solve the equations is of just the same nature as that which one would get with an infinite renormalization of the mass.

I might specify this a little more closely by saying that this infinity is of the form

$$\int_0^\infty d\nu, \tag{40}$$

where ν is the frequency of the emitted photon, when one takes into account the perturbation term in the Hamiltonian that produces the creation of an electron-positron pair and of a photon. With all photon frequencies up to infinity allowed, the integral (40) becomes infinite. One can trace an analogy between this infinity and the infinite rest-mass of an electron in the classical Lorentz theory for a point electron with a Coulomb field around it. The Coulomb field contributes an energy, and when one integrates that energy, taking the electron to be a charge concentrated at a point, one gets an infinity that is essentially of the same nature as (40).

The situation is somewhat modified when one takes into account the complete electron theory, the theory where we have positrons as well as electrons, the positrons appearing as holes in a sea of negative energy electrons. One gets then a different kind of infinity. The integral in (40) gets replaced by a quantity proportional to

$$\int^\infty \frac{d\nu}{\nu}, \tag{41}$$

giving a logarithmic sort of infinity for large ν

$$\sim (\log \nu)_{\nu \to \infty}. \tag{42}$$

This means that the complete theory of electrons does make the infinity less severe.

That can be understood physically in this way. The presence of a charge now produces a polarization in the vacuum around it, because it tends to produce electron pairs there. This, in turn, will to some extent compensate for the Coulomb field of the original electron. It is just this (partial) compensation that makes the infinity (41) less severe than that of integral (40). Still, it is an infinity.

In spite of these difficulties, people have gone on to extensive calculations on these lines. They have calculated how the energy of this electron, connected with the emission operator $\eta^{(e)}$, is affected if there is an external electric field or an external magnetic field. It was found that either one of these fields produces a small correction in the energy of the electron (you have to subtract the infinite term, of course). This small correction is interpreted as giving the Lamb shift in the case of the energy levels of hydrogen or an extra magnetic moment of the electron, the anomalous magnetic moment, for an electron in a magnetic field. These calculations do give results in agreement with observation.

Hence most physicists are very satisfied with the situation. They say: "Quantum electrodynamics is a good theory, and we do not have to worry about it any more." I must say that I am very dissatisfied with the situation, because this so-called "good theory" does involve neglecting infinities which appear in its equations, neglecting them in an arbitrary way. This is just not sensible mathematics. Sensible mathematics involves neglecting a quantity when it turns out to be small—not neglecting it just because it is infinitely great and you do not want it!

One can put the calculations of the Lamb shift and of the anomalous magnetic moment of an electron into a sensible form by introducing a cutoff, by taking the upper integration limit in our integrals to be not infinite but some finite value. The interaction between the electron and the electromagnetic field is then cutoff for frequencies beyond a certain limit (ν_{max}). It is reasonable to take this cutoff frequency to correspond to an energy, say, somewhere around a thousand million volts.

Owing to the appearance of the logarithmic function in (42), the corresponding expression

$$\int^{\nu_{max}} \frac{d\nu}{\nu} \sim \log \nu_{max}$$

with the cutoff will not give appreciably different results. One still gets effectively the same Lamb shifts and the same anomalous magnetic moment when one works with this cutoff, to the first order of accuracy. One then has a theory where the infinities are gone, a theory that is sensible mathematically.

An unfortunate result is that, of course, the relativistic invariance of the theory is spoiled. For, if you have any cutoff at all, thus saying that ν must not exceed a certain value, you are bringing in a non relativistic condition and spoiling the relativistic invariance of the theory. One can thus make quantum electrodynamics into a sensible mathematical theory, but only at the expense of spoiling its relativistic invariance. I think, however, that that is a lesser evil than departing from standard rules of mathematics and neglecting infinite quantities.

I disagree with most physicists at the present time just on this point. I cannot tolerate departing from the standard rules of mathematics. Of course, the proper inference from this work is that the basic equations are not right. There must be some drastic change introduced into them so that no infinities occur in the theory at all and so that we can carry out the solution of the equations sensibly, according to ordinary rules and without being bothered by difficulties. This requirement will necessitate some really drastic changes: simple changes will not do, just because the Heisenberg equations of motion in the present theory are all so satisfactory. I feel that the change required will be just about as drastic as the passage from the Bohr orbit theory to the quantum mechanics.

3 — Magnetic Monopoles

Lecture presented at the Department of Physics, University of Canterbury, Christchurch, New Zealand, September 12, 1975 and organized by Professor B. G. Wybourne, Acting Head of the Department.

I would like to talk to you about one of the applications of quantum theory: the application to magnetic monopoles. I will tell you first about the theoretical ideas which led one to think of monopoles and then I shall tell you about the recent experimental claims to have discovered a monopole.

One works from the idea of the Schrödinger wave function. One does not need to have a relativistic treatment for this argument; one can discuss it just from the point of view of three-dimensional space. One then has a wave function (for some particle), ψ, say, which is a function of the three coordinate x_1, x_2, x_3 and may also vary with the time:

$$\psi(x_1, x_2, x_3; t). \tag{1}$$

One then knows that the usual interpretation of that wave function is that if it is normalized, the square of its modulus $|\psi|^2$ gives us the probability of the particle being in any particular place.

Now, this ψ is usually a complex number, and we are able to multiply it by a *phase factor*. A phase factor is a number of the form $e^{i\gamma}$, where γ is a real number, so that $e^{i\gamma}$ is a number of modulus unity. Hence if we multiply ψ by $e^{i\gamma}$, we get another wave function, Ψ, say

$$\Psi \equiv e^{i\gamma}\psi \tag{2}$$

having its modulus squared the same as that of ψ:

$$|\Psi|^2 = |\psi|^2, \tag{3}$$

so that Ψ corresponds to the same probability distribution as that defined by ψ.

Now, γ does not have to be just a number in (2): it could be a function of position and also even of time. We therefore take γ here to be a function of x_1, x_2, x_3 and also of t. We then still have our new Ψ corresponding to the same probability distribution as that of ψ. We have

$$\Psi(x_1, x_2, x_3; t) = e^{i\gamma(x_1, x_2, x_3; t)}\psi(x_1, x_2, x_3; t). \tag{4}$$

This new Ψ and the original ψ do not, however, satisfy the same wave equation. For, if we form (corresponding to ip_r) $\partial\Psi/\partial x_r$, r taking on the values 1, 2 or 3, we have

$$\frac{\partial\Psi}{\partial x_r} = e^{i\gamma}\left(\frac{\partial}{\partial x_r} + iK_r\right)\psi, \tag{5}$$

where K_r is a function of position:

$$K_r \equiv \frac{\partial\gamma}{\partial x_r}. \tag{6}$$

If we have ψ satisfying some wave equation involving the $\partial/\partial x_r$ of ψ, the function Ψ will satisfy the corresponding wave equation with the partial derivatives $\partial/\partial x_r$ replaced by the operators $\partial/\partial x_r + iK_r$.

Now, let us go one step further. Let us suppose that γ now becomes what mathematicians call a nonintegrable function. Let us imagine that γ does not have a definite numerical value at any point but that it still has a definite change in value when one goes from one point to a neighboring point. We may continue to move our point and so go on around a closed loop. With this motion γ will be continually changing and, as the result of all these changes, it may end up (at the return to the starting point of the loop) with a different value from the one it had at the beginning. Thus the value of γ may change when we go around a closed loop, and for that reason, γ does not have a specified value at each point.

Let us now introduce such a (nonintegrable) γ into the phase factor in expression (4). The result would then be that we would still have an equation with operators of the type of those occurring in (5) applying. But K_r in it would now refer to the change in γ when we go from one point to a neighboring point, and that change would no longer be expressible as a gradient of a scalar [that is to say, expression (6) would no longer apply]. We would have to consider K_r as something more general, something such that when we take $K_r\,dx_r$ and integrate around a closed loop, the result need not be zero:

$$\oint K_r\,dx_r\ \textit{need not be equal to}\ 0. \tag{7}$$

If we do that, we get a physical theory which is definitely more general than what we had before.

It is not, however, essentially new, because it is very similar to the equation which we have for an electron in an electromagnetic field. If we have the theory of an electron in the absence of any field (it could be either the classical, Hamiltonian theory or the quantum theory), we can introduce the field by taking the momentum variables, p_r, in the absence of the field, and replacing them by $p_r + e/c\,A_r$

$$p_r \rightarrow p_r + \frac{e}{c}\,A_r, \tag{8}$$

the A_r representing the potentials of the field. If then the wave function ψ satisfies a particular wave equation in the absence of any field, then ψ in the presence of the field specified by the potentials A_r will satisfy the corresponding wave equation in which p_r has been replaced by $p_r + e/c\, A_r$, or, since

$$p_r = -i\hbar \frac{\partial}{\partial x_r} \tag{9}$$

the $\partial/\partial x_r$ have been replaced according to

$$\frac{\partial}{\partial x_r} \to \frac{\partial}{\partial x_r} + \frac{ie}{\hbar c}\, A_r. \tag{10}$$

This is the change in the wave equation which corresponds to the introduction of the electromagnetic potentials A_r, and we see that it is of the same character as the change

$$\frac{\partial}{\partial x_r} \to \frac{\partial}{\partial x_r} + iK_r \tag{11}$$

introduced by the nonintegrable phase factor above. These changes will be, in fact, the same if

$$K_r = \frac{e}{\hbar c}\, A_r. \tag{12}$$

What that means is that introducing a nonintegrable phase is the same as introducing electromagnetic potentials, the potentials being connected with the nonintegrability of the phase by equation (12). This then gives us a new picture for the electromagnetic potentials. It does not give us, at this stage, any new physical theory. It is simply a new mathematical picture for the Schrödinger equation in which there are potentials acting on the electron.

Let us now consider the total change in phase, that is to say, the total change in γ, when we go around a closed loop:

$$(\Delta\gamma)_{\text{loop}} = \oint_{\text{loop}} K_r\, dx_r. \tag{13}$$

If we now identify K_r with $e/\hbar c\, A_r$ [according to (12)], we have

$$(\Delta\gamma)_{\text{loop}} = \frac{e}{\hbar c} \oint_{\text{loop}} A_r\, dx_r.$$

We may now use Stokes' theorem, which enables us to express any integral taken around a loop as a surface integral taken over a surface that has this loop for its perimeter. We get

$$(\Delta\gamma)_{\text{loop}} = \frac{e}{\hbar c} \iint_{\substack{\text{surface} \\ \text{capping the loop}}} (\text{curl } \mathbf{A}) \cdot d\mathbf{S}. \tag{14}$$

We then get:

$$(\Delta\gamma)_{\text{loop}} = \frac{e}{\hbar c} \iint_{\substack{\text{surface} \\ \text{capping the loop}}} \mathcal{H} \cdot d\mathbf{S}, \tag{15}$$

$\mathcal{H}$ being the magnetic field. Thus we have the numerical factor $e/\hbar c$ multiplied into the magnetic flux going through the loop is equal to the total change in γ when we go round the loop.

Now we have to bring in a new factor: we must take into account that γ, purely as a phaselike factor in the wave function, is undetermined to the extent that we can add to it any integral multiple of 2π. In fact, if we replace γ by $\gamma + 2\pi$ in (2), this equation is not affected at all. If we then take the equation (15) which we have just deduced, we see that its left-hand side is really undetermined to the extent that we can add to it any integral multiple of 2π. Thus the equation cannot be a complete and definite equation as it stands; the left-hand side is undetermined, and the right-hand side is completely definite. Thus, with this picture for interpreting the electromagnetic potentials in terms of a nonintegrable phase, we will have to generalize equation (15) to read:

$$(\Delta\gamma)_{\text{loop}} + 2\pi n = \frac{e}{\hbar c} \iint_{\substack{\text{surface} \\ \text{capping the loop}}} \mathcal{H} \cdot d\mathbf{S}, \tag{16}$$

n being an integer, positive or negative. So, if we take any closed loop, the change in γ when we go around that closed loop plus an unknown integral multiple of 2π equals $(e/\hbar c)$ times the magnetic flux going through the loop.

Let me also bring to your attention the fact that, while I have been considering here just one particular wave function, the same nonintegrable phase factor must be associated with all the wave functions. This must be so, in order that the superposition relations between wave functions can be maintained. For, as one can add two wave functions to get another one, so if we multiply one of our wave functions by the phase-factor $e^{i\gamma}$, we must multiply them all by the same factor in order that the superposition relations be not disturbed.

Now, let us go back to equation (16) and apply considerations of continuity. Let us take a small loop first. As we go around it, the change in γ will be only small, because the (point-to-point) change in γ is small, and thus γ cannot change very much going around a small loop. Similarly, the magnetic flux going through a small loop will be small, under normal physical conditions. Now, n has to be an integer and if both $(\Delta\gamma)_{loop}$ and $\iint \mathcal{H} \cdot d\mathbf{S}$ are small, n must be zero. Thus conditions of continuity will usually tell us that n has to be zero.

There is, however, an exceptional case. I said that the change in γ taken around a small loop must be small, but this is not true under certain conditions. If ψ is zero, then γ is completely undetermined, and if ψ is close to zero, then small changes in ψ may correspond to quite an appreciable change in γ. Just take the example in two dimensions of

$$\psi = x_1 + ix_2. \tag{17}$$

Here is a ψ which is zero when x_1 and x_2 are both zero. It is also perfectly continuous in the neighborhood of this point. Now, the change in the phase of this ψ is 2π, when we make a passage around a small loop enclosing the origin $x_1 = x_2 = 0$.

So we have to consider those regions where ψ vanishes. If ψ vanishes, that involves two conditions, and these two conditions would usually be satisfied along a line. We call that a *nodal line*. There are nodal lines, therefore, where ψ vanishes, and if we take a small loop around a nodal line, the change in ψ, when we go around that small loop does not have to be small. It can be 2π or any integral multiple of 2π, even though ψ is perfectly continuous. That is shown by the foregoing example with ψ given by (17). We can thus infer that in equation (16) n is zero for a small loop except when the loop encircles a nodal line.

If we now go over to large loops and apply this formalism to a large loop, we see that the magnetic flux passing through a large loop, multiplied by $e/\hbar c$, will equal the change in γ going round that large loop, plus a contribution of the form $2\pi n$ coming from each nodal line that passes through that large loop. This large loop will enclose some surface, and there will be a contribution of the form $2\pi n$ from each nodal line that cuts through that surface.

Now, let us next apply (16) to a closed surface. A closed surface has no boundary perimeter at all. Thus if we apply (16) to a closed surface, the change in γ in going round the perimeter is zero, because the perimeter itself has shrunk up to zero. We get then the result that $e/\hbar c$

times the magnetic flux crossing the (closed) surface will be equal to the sum of terms of the form $2\pi n$, one such term for each nodal line that passes through the closed surface. If we have a nodal line coming from infinity, cutting through the surface, running inside the surface, and then out again, that nodal line will make two contributions which will just cancel each other. We get a nonzero total only if we have one or more nodal lines that end inside the closed surface.

Thus we are led to the consideration of the following situation. We have some wave functions which involve a nodal line that has an end to it. The end of the nodal line will then be some sort of singularity in the field (we need not discuss this in detail here). If we then take a closed surface surrounding this singularity, the total magnetic flux going through this closed surface, when multiplied by $e/\hbar c$, will be equal to $2\pi n$

$$\frac{e}{\hbar c} \oiint \mathscr{H} \cdot d\mathbf{S} = 2\pi n. \tag{18}$$

Now, if we have a magnetic flux crossing a closed surface, it means that there is some magnetic monopole inside the surface. If we call the strength of that monopole μ, we should have

$$\oiint \mathscr{H} \cdot d\mathbf{S} = 4\pi\mu, \tag{19}$$

as a magnetic analogue of the Gauss theorem of electrostatics. It says that the magnetic flux crossing a closed surface, enclosing a magnetic monopole of strength μ, is equal to 4π times that enclosed pole strength. If we now compare (19) with the result (18) for flux crossing a closed surface enclosing the end of a nodal line we obtain

$$\mu = \frac{\hbar c}{2e} n \tag{20}$$

as the expression for the strength of the magnetic monopole.

This expression comes out quite definitely from the quantum considerations, and there is no escape from it, if we are going to have magnetic monopoles occurring in the quantum theory. Putting it more theoretically, one can say that, in order that we may have an electron moving in a field of a monopole in accordance with the Schrödinger equation, the monopole must have a strength given by the expression (20), otherwise the equations are not consistent. This is demanded by the Schrödinger formalism.

We may use the experimental result that

$$\frac{\hbar c}{e^2} \approx 137 \tag{21}$$

and so get, from (20)

$$\mu \approx \frac{137}{2} en. \tag{22}$$

If we take the smallest nonzero value of n, n equal to one, we get for the minimum strength of the pole

$$\mu_{\min} \approx \frac{137}{2} e, \tag{23}$$

something that is much larger than the charge on the electron. The minimum monopole is thus quite a large thing.

Now, this theory just shows that such monopoles can exist consistently with the Schrödinger equation. There is nothing in it, however, to say that these monopoles have to exist. Whether they exist or not can only be decided by experiment.

There is one argument in favor of the existence of these monopoles, namely, that they would provide an explanation for why electric charge is always quantized. With all the particles observed in Nature, the electric charge is an integral multiple, positive or negative, of e (the charge on the electron). Now, why should that be so? Why could there not be quite different values for the charge occurring for certain particles?

Well, there is no theoretical explanation for this fact, except for this theory of magnetic monopoles. If there exists even a single monopole anywhere, then, in order that a charged particle can interact with it (which means that we can set up a wave equation for the interaction of this charged particle and the monopole which shall be consistent) it is necessary that the charge on that particle and the strength of the monopole should be connected by the relationship (20). Thus if there exists a monopole anywhere, it should be necessary for all the charged particles in Nature to have their charge quantized. This would be a satisfactory thing, because it would explain a feature in Nature for which there is no other explanation known. However, it is not sufficient to prove that monopoles must exist.

Let us now examine how a monopole would appear experimentally. Suppose that we have a monopole here sitting on a particle. It must be quite stable, for there is a conservation of the magnetic pole strength just like the conservation of the electric charge. The Maxwell equations are symmetrical between the electric and magnetic fields, and the Maxwell equations demand the conservation of the electric charge. They would similarly demand the conservation of the magnetic pole strength. This particle which has a monopole sitting on it may, perhaps, not be a stable particle by itself. If it disintegrates, however, there must be some monopole among the products of the disintegration. The monopole itself is something which is quite permanent and cannot disappear. Once you have a monopole here, the only way in which you could make it disappear would be to have another monopole of equal size and opposite sign and have the two interacting with each other. Then they may annihilate each other with the energy going off into some different form. One monopole by itself, however, is perfectly stable, and only two monopoles of opposite signs can disintegrate each other.

Now, if you are going to make monopoles with some high-energy atomic apparatus, you will have to make them simultaneously in pairs, one positive and one negative. People have searched for monopoles with the high-energy machines, and they have not found any. This is not a proof, of course, that the monopoles do not exist, because it could very well be that simply the rest energy of the monopole is too large for a pair of monopoles to be created with the existing machines. We might expect their rest energy to be rather large, because the monopole strength is large, much larger, in fact, than the charge on the electron. It is not very surprising, therefore, that monopoles have never been detected in existing high-energy apparatus.

One might hope to detect them among the cosmic rays. The cosmic rays, coming in from outer space, have energies enormously greater than any energies that can be (at present) created in the laboratory. It could be, then, that there exist monopoles among the cosmic rays, and so people have been searching for them there. They have been searching now for some decades, and it is only recently that claims have been made to have discovered one of them.

"How would the monopole appear experimentally?" one might ask. It will help us if we know what sort of ionization we would expect to be produced if we had a monopole passing with a high velocity through matter. Let us then compare the ionization trail of a monopole with the ionization trail of a charged particle. In Figure 1 we have a moving charged particle with some matter around it. Consider a typical atom *A*.

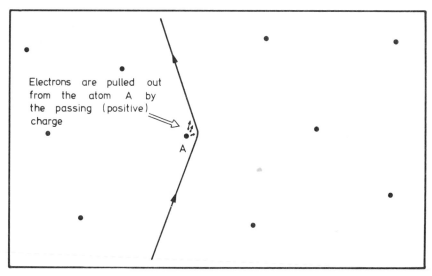

Figure 1. A charged particle (shown here is the case of a positively charged particle) passing through matter causes a trail of ionization being formed along its path.

This charged particle will disturb the electrons of the atom at A, and it may pull some of them out and so produce ionization. The electric force which is pulling the electrons out is then proportional to the charge on the particle:

$$F :: Ze \tag{24}$$

and the time during which that force acts is inversely proportional to the speed v of the particle:

$$t :: v^{-1}. \tag{25}$$

Thus the impulse, which the charged particle passing by gives to one of these electrons, is proportional to the charge and inversely proportional to the speed of the particle:

$$\text{Impulse} = Ft :: Zev^{-1}, \tag{26}$$

that is to say, the faster the particle moves, the less the impulse will be. It will tend to $:: Ze/c$ for particles moving at close to the speed of light. As the particle loses energy and gets toward the end of its trail, the ionization increases, because v gets smaller. Thus we shall have an ionization track which gets thicker as the charged particle gets toward the end of its path.

Now, how is it if we have a monopole going by, instead of a charged particle? The electric field produced by the monopole is proportional to the speed of the monopole. That corresponds to the well-known result that the magnetic field produced by a moving charge is proportional to the speed of the moving charge. Thus with a monopole going by, we have a force proportional to the speed of the monopole:

$$F :: \mu v. \qquad (27)$$

Then the impulse which the monopole gives to one of the electrons here (i.e., one of the electrons on a typical atom A of the surrounding matter of Figure 1) will be the force multiplied by the time and thus will be nearly independent of the speed:

$$\text{Impulse} :: \mu \qquad (28)$$

(and only proportional to the strength μ of the monopole). Thus a monopole going by will produce a trail of ionization which is pretty steady and does not increase toward the end of the trail. That is a way of distinguishing a monopole from an ordinary charged particle.

This is the method which was used by Price and others for their recent experimentation which led them to believe that they had discovered a monopole. What they did was to send some recording apparatus up in a balloon, keep it near the top of the atmosphere for a few days, and then bring down their apparatus and analyze the records. The apparatus consisted mainly of a pile of Lexan sheets, Lexan being a certain kind of transparent material. There was a whole set of these sheets, one piled on top of the other. Any ionizing particle passing through these sheets damages them in a certain way, and the amount of the damage can be assessed by etching the sheets. The sheets were all etched and the etch marks examined to see how much ionization there was. Figure 2 is a diagram showing the results they obtained. There is a gap at the top, between the first of these Lexan sheets and the remaining ones, and in that gap there was some other apparatus, a Cerenkov detector and also an ordinary emulsion plate. As a result of the observations of this emulsion plate, they were able to infer that the particle was moving downward (just from the delta rays), and with the Cerenkov detector they had some information about the speed of the particle. Figure 2 shows the gap between the top Lexan sheet and the whole remaining stack of 32 sheets. Now, these dots [circles (filled)] and triangles (open) indicate the amount of ionization in the sheets. The

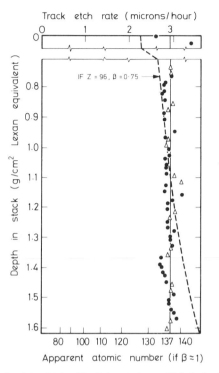

Figure 2. Experimental points obtained by Price et al. on which their claim of having possibly discovered a monopole (details in text) is based.

circles and the triangles differ only through having different times of exposure to the acid that caused the etching. These "points" [so marked by the circles and the triangles] do lie approximately along a vertical line, indicating that the amount of ionization is fairly constant.

Now, if there was a charged particle instead of a monopole producing this ionization, the curve (through the experimental points) should go toward the right as one proceeds downwards, because the ionization should increase toward the end of the trail. The dashed line is what one gets with a charge of 96*e* and a speed of three quarters of the speed of light, and you see that this sloping curve does not fit the data at all. The vertical line, on the other hand, does fit the data fairly well, even though there are quite a lot of discrepancies, experimental errors, here. This figure supplied the basis for the claim by Price and his co-workers that it was a magnetic monopole which they had caught in their apparatus.

We were discussing this work with the physicists in Sydney and someone there said that one possible cause of error would have been if there maybe was some saturation phenomenon coming in with the etching process. Thus if the plate was very heavily damaged by ionization there would not be the corresponding increase in the size of the etch marks, owing to some saturation effect. That was a possible criticism, and I do not know to what extent it is valid. This was, after all, only a preliminary report that was issued (and from which Figure 2 was taken) and, of course, this criticism is something that will have to be looked into very carefully, to see whether the size of the etch mark does really correspond to the amount of ionization the particle is producing.

Professor George in Sydney was sufficiently interested to ring up Alvarez in Berkeley (who is the head of the labroatory where this work was done) and to ask Alvarez what was his opinion about it. Alvarez was very hostile to the interpretation of Price and the others. He said that it could be that there was some charged particle producing the ionization corresponding to the etchings down to a certain depth and that, at that depth, the particle hit some nucleus and underwent a disintegration, as a result of which it then continued with a reduced charge.

On the basis of this telephone conversation with Alvarez, the physicists in Sydney have constructed Figure 3. They suppose here that you start off with a charged particle (at the point A, say), which they took to have $Z = 96$. It then continued up to the point B, and there it hit an atomic nucleus. It then underwent a disintegration so that it had a reduced charge (point C) and continued down from C along the curve CD. Maybe this new particle was itself unstable, or, perhaps, it met another particle at D. It then underwent another transformation there, shedding some more charge. It could be that this would provide an explanation alternative to the monopole explanation.

This is a question which has not yet been resolved. In any case, if Alvarez is right, it would be a sort of a coincidence that the overall picture yields a line that is very closely vertical. It would seem that Nature has been trying to deceive us. I do not know what the real answer is, and we have to wait for the experimental physicists to study these results in greater detail and come to some conclusion.

Professor Hora in Sydney rang up Hofstadter to ask him what his opinion was. Hofstadter was rather noncommittal and said that he

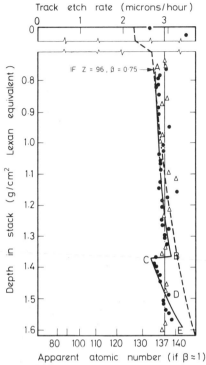

Figure 3. A possible interpretation of experimental points of Price et al. (Fig. 2) in terms of fragmentation processes (see text for details). The particular interpretation shown as a full line, results from calculations by E. P. George (August 1975).

thought it was fifty-fifty whether Price was right or Alvarez was right. I think that that is perhaps my opinion also. We will have to leave the question open for the time being.

There is, however, one type of observation which rather speaks against the interpretation of these results in terms of a monopole. For, if monopoles are coming in the cosmic rays, then presumably lots of them would have arrived in the past, having come down through the atmosphere and stopped somewhere or other where we ought to be able to find them. One monopole by itself is perfectly stable, and if the total number of monopoles [in a region] is rather small, the chance of one monopole meeting another monopole of opposite sign is very small, so that they do not have much chance to annihilate each other. If there are these monopoles coming in the cosmic rays, we would expect to have a lot of them lying around. People have searched for them in many places, all the places they could think of, and they have not found any yet. The monopoles would probably tend to go toward the poles of the Earth's magnetic field, and people have made searches for them at regions of high latitude for that reason. Still, they have not found any.

It may be that the monopoles are sinking through the Earth in the hope of getting close to the Earth's magnetic poles. It could be that they just have not been found because they penetrate too far into the Earth's surface. We do not really know how well they should penetrate through solid matter. I do not think anyone has estimated that. It would be pretty difficult to estimate without knowing more about the monopoles. Further, Price and his co-workers believe that their monopole has a mass of some two hundred proton masses or even larger. Of course, if that is the case, it would explain why monopoles are not observed with the high-energy machines. Those machine would not have enough energy (at present) to create a pair of particles with this mass: we would need a simply enormous energy in the rest-energy frame of the collision process.

I think I must conclude at this point and leave it as a question for you to puzzle over whether monopoles exist or not. I hope that the experimentalists will come up with something more definite (Price and the other workers have promised another paper. This one was published in *Physical Review Letters* of 25th August).

THE PRESENT STATE OF THE EXPERIMENTAL EVIDENCE FOR THE MONOPOLE (DECEMBER 1976)

A monopole is indestructible unless it meets another monopole of opposite sign. We can be as certain of this as of the conservation of electricity. Both statements come from general properties of the Maxwell equations.

If monopoles are raining down on the earth from outer space, they must be lying about somewhere. Very extensive searches have been made for them, with negative results. It could be that they have penetrated to great depths in the earth. But they are easily trapped by ferromagnetic materials; thus we should expect to find some near the surface.

Because of the great weight of this negative evidence, most physicists do not believe that there can be monopoles in the cosmic rays, and they seek for some alternative explanation of the event that Price and his co-workers discovered. This is not easy. The original explanation of Alvarez, that it was a platinum nucleus that underwent fragmentation on its way through the Lexan plates, is disproved by the etching of some further plates that were not included in the first batch.

It seems that no known particle will fit the observations. If it is a charged nucleus, it would have to be a superheavy one, beyond the range of all the known ones. Alternatively, it might be a nucleus of antimatter.

It is very difficult to decide about a particle when there is only one example of it. One very much hopes that further examples will be discovered in future experiments.

4 — A Positive–Energy Relativistic Wave Equation

Lecture presented at the Department of Theoretical Physics, University of New South Wales, Kensington, Sydney, Australia, August 26, 1975, and organized by Professor H. Hora, Head of the Department.

I would like to tell you today about an interesting development in relativistic quantum mechanics, the development of a new relativistic wave equation. I will tell you then about some basic characteristics of that equation and discuss the nature of its solutions. The new equation is a relativistic wave equation, and it describes certain particles with nonzero rest mass. It will be found to bear some formal similarity to the familiar relativistic wave equation, which describes particles that have spin angular momentum of half a quantum. That similarity, however, is only superficial, and the particles described by the new equation would, if they exist, have characteristics quite different from those of the spin one-half particles (like electrons, protons, and so on) described by that older equation.

Let us remind ourselves briefly about the form of that older equation. For the sake of simplicity and for particles of nonzero rest mass, m, say, we shall choose units in which

$$\hbar = m = c = 1 \tag{1}$$

where $\hbar$ and c have their usual significance. In those units, that older relativistic wave equation has the form

$$\left\{ \frac{\partial}{\partial x_0} + \alpha_r \frac{\partial}{\partial x_r} + i\alpha_m \right\} \psi = 0 \tag{2}$$

in which the usual summation convention has been used, so that $\alpha_r(\partial/\partial x_r)$ stands for the sum: $\alpha_1(\partial/\partial x_1) + \alpha_2(\partial/\partial x_2) + \alpha_3(\partial/\partial x_3)$, since the index r has here the range $r = 1, 2, 3$. The symbols x_0, x_1, x_2 and x_3 stand, respectively, for t, x, y, and z, the four space-time coordinates of the particle. The coefficients α_r, with $r = 1$, 2, and 3, and α_m are certain four-by-four matrices which must be chosen so that they all anticommute with one another and so that the square of each of them is equal to the unit (four-by-four) matrix. This means that we have

$$[\alpha_r, \alpha_s]_+ = \alpha_r \alpha_s + \alpha_s \alpha_r = 0 \tag{3}$$

for $r \neq s$ with $r, s = 1$, 2, and 3

$$(\alpha_1)^2 = (\alpha_2)^2 = (\alpha_3)^2 = (\alpha_m)^2 = 1 \tag{4}$$

and

$$[\alpha_r, \alpha_m]_+ = 0, \tag{5}$$

The quantity ψ in (2) is a four-rowed single-column matrix whose elements are each some function of the coordinates x_0, x_1, x_2, and x_3. That ψ plays the role analogous to that played in the usual Schrödinger equation, by the conventional single-component wave function. The

equation (2) has been quite successful in describing spin-one-half particles, such as electrons, in relativistically covenant quantum terms.

Now, in the development of the new equation, it is assumed that the
particle described by it is not structureless, as was the case with the
spin-one-half particles described by (2), but that it has a certain dynamic
structure; that is to say, that it has some internal degrees of freedom.
[Incidentally, in his well-known book *The Principles of Quantum Mechanics* Professor Dirac remarks that in the case of equation (2), the α_r,
α_m matrices "describe some new degrees of freedom, belonging to some
internal motion" and he states that "they bring in the spin" of the
particle. This follows, because those matrices are independent of the
coordinates x_0, x_r and "so they must commute with" the momenta and
the coordinates.]

These new degrees of freedom are to be associated here with certain
dynamical variables (q_1, p_1) and (q_2, p_2), to be thought of as corresponding to two independent linear harmonic oscillators. This then
leads to the following commutation relations:

$$\left.\begin{aligned}
q_1 q_2 - q_2 q_1 &\equiv [q_1, q_2]_- = [p_1, p_2]_- = 0 \\
[q_1, p_1]_- &= [q_2, p_2]_- = i \\
[q_1, p_2]_- &= [q_2, p_1]_- = 0
\end{aligned}\right\} \tag{6}$$

among these dynamical variables. These relations may be written in a
particularly convenient form if we introduce the definitions

$$q_3 \equiv p_1, \qquad q_4 \equiv p_2 \tag{7}$$

and then construct a four-by-four matrix, β, such that

$$q_a q_b - q_b q_a \equiv [q_a, q_b]_- = i\beta_{ab}, \tag{8}$$

where

$$a, b = 1, 2, 3, \text{ and } 4. \tag{9}$$

The β matrix, in view of relations (7) and the definition (8), must have
the form

$$\beta = \begin{pmatrix} 0 & 0 & 1 & 0 \\ 0 & 0 & 0 & 1 \\ -1 & 0 & 0 & 0 \\ 0 & -1 & 0 & 0 \end{pmatrix}. \tag{10}$$

The β matrix is clearly antisymmetric, and its square is equal to the negative of the unit matrix:

$$\beta_{ab} = -\beta_{ba}, \quad (\beta)^2 = -1. \tag{11}$$

The above requirement that the new degrees of freedom be associated with dynamical variables of two independent harmonic oscillators may be expressed more formally and precisely. The wave function ψ which corresponds to the particle, in addition to being a function of the particle's coordinates x_0, x_1, x_2, x_3, must also depend on two independent, that is to say, commuting, q_a variables, a being equal to 1, 2, 3 and 4. Such pairs of commuting q_as are

$$(q_1, q_2), (q_3, q_4), (q_1, q_4) \text{ and } (q_2, q_3).$$

We choose one of these pairs, say, (q_1, q_2) and then require the following functional dependence for the wave function:

$$\psi = \psi(x_0, x_1, x_2, x_3; q_1, q_2). \tag{12}$$

A four-rowed column matrix may be next constructed from the q_as and the wave function ψ:

$$q\psi \equiv \begin{pmatrix} q_1\psi \\ q_2\psi \\ q_3\psi \\ q_4\psi \end{pmatrix}, \tag{13}$$

wherein the symbol q stands for the matrix

$$q \equiv \begin{pmatrix} q_1 \\ q_2 \\ q_3 \\ q_4 \end{pmatrix} \tag{14}$$

If we then adopt a line of argument essentially the same as that originally used in deriving the form of equation (2), we may write our new equation as

$$\left\{ \frac{\partial}{\partial x_0} + \alpha_r \frac{\partial}{\partial x_r} + \beta \right\} q\psi = 0, \tag{15}$$

where the α_r coefficients, with $r = 1$, 2, and 3, are again four-by-four matrices chosen so that they anticommute with one another and also with β:

$$\left. \begin{array}{l} [\alpha_r, \alpha_s]_+ = 0, \quad r \neq s \\[2mm] [\alpha_r, \beta]_+ = 0. \end{array} \right\} \tag{16}$$

We are using the notation $[a, b]_+ = ab + ba$. We require also the square of each α_r to be equal to the unit matrix:

$$(\alpha_1)^2 = (\alpha_2)^2 = (\alpha_3)^2 = 1. \qquad (17)$$

The matrix β in our new equation (15) is the same as the β of equation (10). We get an intimate association of the matrices with the new degrees of freedom [see equation (8)].

Now, the mathematical form or our new equation (15) is clearly very similar to that of the older equation [equation (2)]. Its differences from that equation all stem from the association of the new equation with the new degrees of freedom. This causes $q\psi$ to be very different from the ψ in equation (2). Also β is directly connected with the commutation relations among the variables defining those degrees of freedom.

We now impose further restrictions on the α matrices, which will be useful later. We shall assume that they are real and symmetrical (and hence, Hermitean). There are, of course, many ways in which these conditions can be satisfied in addition to the conditions (16) and (17). We take here, by the way of an example, the following set of α_rs:

$$\alpha_1 = \begin{pmatrix} 0 & 0 & -1 & 0 \\ 0 & 0 & 0 & 1 \\ -1 & 0 & 0 & 0 \\ 0 & 1 & 0 & 0 \end{pmatrix}, \quad \alpha_2 = \begin{pmatrix} 0 & 0 & 0 & 1 \\ 0 & 0 & 1 & 0 \\ 0 & 1 & 0 & 0 \\ 1 & 0 & 0 & 0 \end{pmatrix}, \quad \alpha_3 = \begin{pmatrix} 1 & 0 & 0 & 0 \\ 0 & 1 & 0 & 0 \\ 0 & 0 & -1 & 0 \\ 0 & 0 & 0 & -1 \end{pmatrix}.$$

$$(18)$$

If we now substitute this set into equation (15) and note that

$$\alpha_1 q = \begin{pmatrix} -q_3 \\ q_1 \\ -q_1 \\ q_2 \end{pmatrix}, \quad \alpha_2 q = \begin{pmatrix} q_4 \\ q_3 \\ q_2 \\ q_1 \end{pmatrix}, \quad \alpha_3 q = \begin{pmatrix} q_1 \\ q_2 \\ -q_3 \\ -q_4 \end{pmatrix}, \quad \beta q = \begin{pmatrix} q_3 \\ q_4 \\ -q_1 \\ -q_2 \end{pmatrix}. \qquad (19)$$

we obtain the following set of four equations:

$$\{(\partial^0 + \partial^3)q_1 - \partial^1 q_3 + \partial^2 q_4 + q_3\}\psi = 0 \quad (a)$$
$$\{(\partial^0 + \partial^3)q_2 + \partial^1 q_4 + \partial^2 q_3 + q_4\}\psi = 0 \quad (b)$$
$$\{(\partial^0 - \partial^3)q_3 - \partial^1 q_1 + \partial^2 q_2 - q_1\}\psi = 0 \quad (c) \qquad (20)$$
$$\{(\partial^0 - \partial^3)q_4 + \partial^1 q_2 + \partial^2 q_1 - q_2\}\psi = 0, \quad (d)$$

where we have used the abbreviated notation

$$\partial^\mu \equiv \frac{\partial}{\partial x_\mu}, \quad \mu = 0, 1, 2, \text{ and } 3. \qquad (21)$$

In the representation we are using for the two oscillators, we interpret $q_1\psi$ and $q_2\psi$ as just the numbers obtained from multiplying ψ, respectively, by the numbers q_1 and q_2 but $q_3\psi$ and $q_4\psi$ as

$$q_3\psi = -i\frac{\partial}{\partial q_1}\psi, \qquad q_4\psi = -i\frac{\partial}{\partial q_2}\psi.$$

Since, as we see from set (20), our equation (15) is really four equations, all four of them referring to the same ψ so that we have four *linear* equations for only one unknown. There arises the question of consistency. First of all, they are not all independent. This can be seen at once if we form the sum: q_2 times (20a) plus $- q_1$ times (20b) plus $- q_4$ times (20c) plus q_3 times (20d). Then we have, in view of the commutation relations (8) and since the q_as are independent of the x_μ coordinates, the left-hand side of the sum is equal to zero so that we simply have an identity $0 = 0$. Only three equations in the set (20) can, therefore, be independent of each other. It is important now to check the internal consistency of the three equations.

Let us write the equations (20) as

$$P_a\psi = 0, \qquad a = 1, 2, 3, \text{ and } 4 \tag{22}$$

where P is the four-row single-column matrix

$$P \equiv (\partial^0 + \alpha_r\partial^r + \beta)q, \tag{23}$$

with elements

$$P_a = (\partial^0 + \alpha_r\partial^r + \beta)_{ab}q_b. \tag{24}$$

Then the consistency condition will be expressed simply as

$$[P_a, P_b]\psi = 0, \qquad a, b, = 1, 2, 3, \text{ and } 4, \tag{25}$$

which must be satisfied if both $P_a\psi = 0$ and $P_b\psi = 0$, that is, if the same ψ satisfies all the equations (20). We have, however, that

$$
\begin{aligned}
[P_a, P_b]_- &= [(\partial^0 + \alpha_r\partial^r + \beta)_{ac}q_c, (\partial^0 + \alpha_s\partial^s + \beta)_{bd}q_d]_- \\
&= (\partial^0 + \alpha_r\partial^r + \beta)_{ac}(\partial^0 + \alpha_s\partial^s + \beta)_{bd}[q_c, q_d]_- \\
&= (\partial^0 + \alpha_r\partial^r + \beta)_{ac}(\partial^0 + \alpha_s\partial^s + \beta)_{bd}i\beta_{cd} \\
&= i(\partial^0 + \alpha_r\partial^r + \beta)_{ac}\beta_{cd}(\partial^0 + \alpha_s\partial^s - \beta)_{db} \\
&= i\{(\partial^0 + \alpha_r\partial^r + \beta)\beta(\partial^0 + \alpha_s\partial^s - \beta)\}_{ab} \\
&= i\{(\partial^0 + \alpha_r\partial^r + \beta)(\partial^0 - \alpha_s\partial^s - \beta)\beta\}_{ab} \\
&= i\{(\partial^0\partial^0 - \partial^r\partial^r + 1)\beta\}_{ab},
\end{aligned}
$$

where we have made use of relations (8) and of the symmetry of the α_r and antisymmetry of β, as well as of the relations (16), (17), and (11). It therefore follows that the consistency condition (25) is equivalent to

$$(\partial_\mu \partial^\mu + 1)\psi = 0, \tag{26}$$

or to the De Broglie equation for the wave function ψ for a particle of rest-mass unity. The operator

$$\partial_\mu \partial^\mu = g_{\mu\nu} \partial^\mu \partial^\nu \tag{27}$$

is known as the D'Alembertian operator. Thus, in order that the four differential equations (20) shall be mutually consistent, the wave-function ψ must satisfy the De Broglie equation (26) as well as these equations.

It is required, of course, that the new wave equation be relativistically correct, which means that it should be Lorentz invariant. It is interesting to note that (15) can be, in fact, rewritten in a form which gives it a relativistically covariant appearance. For this purpose, one introduces new matrices:

$$\gamma_\mu \equiv \beta\alpha_\mu, \qquad \mu = 0, 1, 2, \text{ and } 3, \tag{28}$$

where

$$\alpha_0 \equiv 1. \tag{29}$$

Then, on multiplying (15) from the left by β and using (28) and (29), we obtain

$$(\gamma_\mu \partial^\mu - 1)_q \psi = 0 \tag{30}$$

with the γ_μ satisfying the anticommutation relations

$$[\gamma_\mu, \gamma_\nu]_+ = -2g_{\mu\nu}. \tag{31}$$

Now, (30) and (31) certainly *look* relativistically covariant, but they cannot be manifestly relativistic because of the difference between the symmetry character of γ_0 (which is antisymmetric) and the γ_rs (which are symmetric) [since $(\gamma_0)_{ba} = \beta_{ba} = -\beta_{ab} = -(\gamma_0)_{ab}$ and $(\gamma_r)_{ba} = (\beta\alpha_r)_{ba} = \beta_{bc}(\alpha_r)_{ca} = -\beta_{cb}(\alpha_r)_{ac} = -(\alpha_r\beta)_{ab} = (\beta\alpha_r)_{ab} = (\gamma_r)_{ab}$]. Thus the temporal ($\gamma_0$) is distinguished from the spatial (γ_r) members of the γ_μ set.

We need, therefore, to approach the question of the Lorentz invariance of (15) in a more systematic way. Fortunately, the method of testing for invariance is already well known. It starts with the application of an infinitesimal Lorentz transformation to that equation. It then proceeds in step-by-step analogy to the way this was done for verifying the Lorentz invariance of the equation (2). For this reason, I do not enter here into details, but only sketch out the main steps to be followed. The required infinitesimal Lorentz transformation on the coordinates x_μ, with $\mu = 0$, 1, 2, and 3, transforms them into some new coordinates:

$$x_\mu^* = x_\mu + a_\mu^{\ \nu} x_\nu, \tag{32}$$

where the real quantities

$$a_{\mu\nu} \equiv g_{\mu k} a_\nu^{\ k} \tag{33}$$

are infinitesimal and antisymmetric:

$$a_{\nu\mu} = - a_{\mu\nu}. \tag{34}$$

We have, accordingly,

$$\partial^{\mu*} = \partial^\mu + a^\mu_{\ \nu} \partial^\nu = \partial^\mu - a_\nu^{\ \mu} \partial^\nu, \tag{35}$$

and to the first order in small quantities,

$$\partial^\mu = \partial^{\mu*} + a_\nu^{\ \mu} \partial^{\nu*}. \tag{36}$$

Now, this expression (36) may be put into equation (15), which may be rewritten more concisely as

$$(\alpha_\mu \partial^\mu + \beta) q \psi = 0. \tag{37}$$

This substitution then gives, after rearrangement,

$$\{(\alpha_\mu + a_\mu^{\ \nu} \alpha_\nu) \partial^{\mu*} + \beta\} q \psi = 0. \tag{38}$$

If we then introduce, for convenience, a symmetric matrix,

$$N \equiv \tfrac{1}{4} a^{\mu\nu} \alpha_\mu \beta \alpha_\nu \tag{39}$$

it is readily shown that (38) may be rearranged to give

$$(\alpha_\mu \partial^\mu + \beta) q^* \psi = 0, \tag{40}$$

where we have introduced the quantity

$$q^* \equiv (1 - \beta N) q. \tag{41}$$

Since, however, the q^* matrices satisfy the commutation relations

$$[q_a^*, q_b^*]_- = i\beta_{ib}, \tag{42}$$

they will play just the role of q matrices. It therefore follows that the transformation (32), (41) preserves the form of the equation (37) [or

(15)]. This form will thus also be unchanged by any finite proper Lorentz transformation.

It is interesting to note that the transformation (41) of q into $q*$ may be rewritten in the form of an infinitesimal unitary transformation. Indeed, we have to the first order in small quantities:

$$q* = \left(1 - \frac{1}{4}a^{\mu\nu}\beta\alpha_\mu\beta\alpha_\nu\right) \quad q = \left(1 - \frac{i}{2}W\right)q\left(1 + \frac{i}{2}W\right) \tag{43}$$

where W is the quantity

$$W \equiv \tfrac{1}{4}a^{\mu\nu}q^\sim\alpha_\mu\beta\alpha_\nu q. \tag{44}$$

Here we have introduced the notation

$$q^\sim \equiv (q_1, q_2, q_3, q_4) \tag{45}$$

for the single-row matrix that is the transpose of the single-column matrix q. The right-hand side of equation (43) is the result of an infinitesimal unitary transformation, W being an infinitesimal Hermitean operator.

We now come to a very interesting feature of the new wave equation, namely, that it automatically allows only positive energy solutions. To see this, we rewrite the equations (20) in terms of the energy-momentum eigenstates of the particle. We shall write, therefore, temporarily,

$$p^\mu \equiv i\partial^\mu$$

for the energy-momentum operators for the particle and then use the symbols p^μ for the eigenvalues of the p^μ operators, which we shall take to be real numbers. Thus

$$i\partial^\mu\psi = p^\mu\psi. \tag{46}$$

On using the relations (46) in equations (20), we obtain

$$\{(p_0 - p_3)q_1 + (i + p_1)q_3 - p_2q_4\}\psi = 0 \quad \text{(a)}$$

$$\{(p_0 - p_3)q_2 - p_2q_3 + (i - p_1)q_4\}\psi = 0 \quad \text{(b)}$$

$$\{(p_0 + p_3)q_3 - (i - p_1)q_1 - p_2q_2\}\psi = 0 \quad \text{(c)} \tag{47}$$

$$\{(p_0 + p_3)q_4 - p_2q_1 - (i + p_1)q_2\}\psi = 0, \quad \text{(d)}$$

where we have used the relations

$$p^0 = p_0, \quad p^r = -p_r, \quad r = 1, 2, \text{ and } 3. \tag{48}$$

[It should be noted that p_1, p_2, and p_3 symbols used here refer to the momentum components of the particle and must not be confused with the same symbols used earlier, in equations (6) and (7) and the accompanying text, for the momentum variables of the two harmonic oscillators; p_1 corresponding to the first oscillator and being later replaced by the symbol q_3, while p_2 corresponded to the second oscillator and was replaced by the symbol q_4].

Now, through an appropriate Lorentz transformation, we may use the frame in which the particle has zero momentum, so that

$$p_1 = p_2 = p_3 = 0. \tag{49}$$

In that frame equations (47) become

$$
\begin{array}{ll}
(p_0 q_1 + i q_3)\psi = 0 & \text{(a)} \\
(p_0 q_2 + i q_4)\psi = 0 & \text{(b)} \\
(p_0 q_3 - i q_1)\psi = 0 & \text{(c)} \\
(p_0 q_4 - i q_2)\psi = 0 & \text{(d)}.
\end{array}
\tag{50}
$$

The De Broglie equation (26), in terms of the p^μ and p_μ quantities, takes in general the form

$$(p_\mu p^\mu - 1)\psi = 0. \tag{51}$$

Thus, in that special frame, we have

$$(p_0)^2 = 1, \tag{52}$$

This implies that the particle energy p_0 can take only the values of $+1$ or -1, for our rest mass $= 1$ particle. Thus equations (50) simplify to (the top sign corresponding to $p_0 = +1$, the bottom one to $p_0 = -1$):

$$
\begin{array}{ll}
(\pm q_1 + i q_3)\psi = 0 & \text{(a)} \\
(\pm q_2 + i q_4)\psi = 0 & \text{(b)} \\
(\pm q_3 - i q_1)\psi = 0 & \text{(c)} \\
(\pm q_4 - i q_2)\psi = 0 & \text{(d)}.
\end{array}
\tag{53}
$$

It is clear now that, in (53), equations (a) and (c) are equivalent and so are (b) and (d). Thus, in view of the identification (in the q representation where q_1 and q_2 are diagonal),

$$q_3 \equiv -i \frac{\partial}{\partial q_1}, \qquad q_4 \equiv -i \frac{\partial}{\partial q_2}, \tag{54}$$

the wave equations now reduce to only two differential equations

$$\left(q_1 \pm \frac{\partial}{\partial q_1}\right)\psi = 0 \tag{55}$$

and

$$\left(q_2 \pm \frac{\partial}{\partial q_2}\right)\psi = 0 \tag{56}$$

in which the top sign again refers to $p_0 = +1$ and the bottom sign to $p_0 = -1$. These equations integrate then quite simply, giving

$$\psi :: e^{-(q_1^2 + q_2^2)/2} \quad \text{for } p_0 = +1 \tag{57}$$

and

$$\psi :: e^{+(q_1^2 + q_2^2)/2} \quad \text{for } p_0 = -1. \tag{58}$$

Now, the solution (58) is clearly unacceptable physically, because it diverges for large values of q_1 and q_2 and thus cannot be normalized. Only the solution (57) is admissible, and hence only the $p_0 = +1$ energy eigenvalue is allowed. The negative energy eigenvalue is automatically excluded through the impossibility of solution (58).

In frames other than the special one just discussed, the result that only positive energy solutions give physically admissible wave functions remains. [Prof. Dirac states, in his *Proc. Roy. Soc.* paper of 1971, that in the general case, equations (47) integrate to give

$$\exp\left\{-\left[\frac{q_1^2 + q_2^2 + ip_1(q_1^2 - q_2^2) - 2ip_2q_1q_2)}{2(p_0 + p_3)}\right]\right\} e^{-ip^\mu x_\mu},$$

which shows again that the energy p_0 must be positive to yield an acceptable solution.]

The fact that the new wave equations allow only positive energy values for the particle represented by them is very remarkable by itself. This is, however, not an altogether new result; already in 1932 a Lorentz-invariant wave equation which allowed only positive energy solutions had been proposed by Majorana. That equation had the form

$$(q^\sim \alpha_\nu q \partial^\nu + 2i)\psi = 0 \tag{59}$$

and was by itself Lorentz-invariant. Now, if we multiply equation (37) by linear functions of the qs, we obtain equations which have, in general, the form

$$\{q^\sim \lambda \alpha_\mu q \partial^\mu + q^\sim \lambda \beta q\}\psi = 0 \tag{60}$$

where λ is some 4×4 matrix, so that $q^{\sim}\lambda$ is a linear function of qs. It then turns out that there are fifteen independent λ matrices giving different results. Thus there are fifteen different equations of the type (60), quadratic in the qs. One of these equations, namely, that corresponding to $\lambda = 1$, is identical with the Majorana equation (59).

It would be tempting, therefore, to limit our attention to the consideration of the Majorana equation and its solutions. That equation has been studied in detail by Majorana himself, and he was able to show that it leads to a whole mass spectrum with an infinity of mass values. This looks very promising at first, but, when one works out the spin of the Majorana particles, one finds that an increase in the particle spin is accompanied by a decrease in the particle mass, so that more massive Majorana particles would have to possess less spin than the lighter ones. This is, of course, completely contrary to experimental evidence. This fact caused the abandonment of the Majorana equation by physicists.

We must, therefore, keep to a fixed mass value for our particle, thus retaining the De Broglie equation (26). This amounts, however, to keeping the whole set of equations (60), all fifteen of them. Although one of them will have the Majorana form (59), it will, when taken in conjunction with the remaining fourteen, no longer lead to the above-mentioned undesired properties for our particles.

We now come to the question of the particle spin. As there is not much time left, I will deal with it only in outline. The general approach one uses to find out about the spin of a particle described by a wave equation is to apply an infinitesimal rotation (about the origin) to the wave function itself, a rotation containing a term with a spin operator in it, and then require that the transformed wave function should satisfy the same wave equation which the untransformed wave function satisfied. We, therefore, transform our ψ in (37) to the "rotated" wave function

$$(1 + \tfrac{1}{2}a^{\rho\sigma}M_{\rho\sigma})\psi, \tag{61}$$

where the operator $M_{\rho\sigma}$ has the form

$$M_{\rho\sigma} = x_\rho\partial_\sigma - x_\sigma\partial_\rho - is_{\rho\sigma}, \tag{62}$$

the $s_{\rho\sigma}$ representing the spin operators which operate on the qs. On requiring then (61) to satisfy equation (37), one finally obtains the identification

$$s_{\rho\sigma} = -\tfrac{1}{4}q^{\sim}\alpha_\rho\beta\alpha_\sigma q + \tfrac{1}{2}ig_{\rho\sigma} \tag{63}$$

if $s_{\rho\sigma}$ is antisymmetric. With the choice (18) for the αs, we then have

$$s_{01} = \tfrac{1}{4}(q_1^2 - q_2^2 - q_3^2 + q_4^2)$$
$$s_{02} = \tfrac{1}{2}(q_3 q_4 - q_1 q_2)$$
$$s_{03} = \tfrac{1}{2}(q_1 q_3 + q_4 q_2)$$
$$s_{12} = \tfrac{1}{2}(q_2 q_3 - q_1 q_4) \tag{64}$$
$$s_{23} = \tfrac{1}{2}(q_1 q_2 + q_3 q_4)$$
$$s_{31} = \tfrac{1}{4}(q_1^2 - q_2^2 + q_3^2 - q_4^2)$$

The last three equations of the set (64) then give

$$s_{12}^2 + s_{23}^2 + s_{31}^2 = \tfrac{1}{16}(q_1^2 + q_2^2 + q_3^2 + q_4^2)^2 - \tfrac{1}{4}. \tag{65}$$

According to quantum mechanics, however, the magnitude, s, say, of the spin is defined by

$$s(s + 1) = s_{12}^2 + s_{23}^2 + s_{31}^2. \tag{66}$$

Hence the result (65) gives, for the magnitude s of the spin of the particle described by the equation (37), the expression

$$s = \tfrac{1}{4}(q_1^2 + q_2^2 + q_3^2 + q_4^2) - \tfrac{1}{2}. \tag{67}$$

Now the eigenvalues of $\tfrac{1}{2}(q_1^2 + q_3^2)$, like the energy eigenvalues for a harmonic oscillator, have the form $n + \tfrac{1}{2}$, while those of $\tfrac{1}{2}(q_2^2 + q_4^2)$ have, similarly, the form $n' + \tfrac{1}{2}$, where n and n' are positive integers or zero. Hence the eigenvalues of s are of the form

$$\tfrac{1}{2}(n + n'). \tag{68}$$

It turns out, however, that the wave function satisfying (37) must always be an even function of the qs, so that $n + n'$ is always an even number or zero. Hence the spin magnitude must be always an integer or zero. The particle described by the wave equation (37) is, therefore, a particle with integral spin. [Hence, presumably, it is a boson, obeying the Bose–Einstein statistics.]

Incidentally, there seems to be, at the first sight, a difficulty with the spin. It would appear that s is dependent on the particle's momentum. This turns out, however, to be a spurious result. Since the separation of angular momentum into orbital and spin angular momentum parts is not a relativistic procedure, the spin of the particle may be redefined so as to remove its spurious dependence on the momentum.

Let us now very briefly examine the physical picture of the particle described by the above theory. What is, in particular, the type of internal motion the particle undergoes? This sort of question is best answered when we work in the Heisenberg picture, since the Heisenberg picture gives us the type of information closest to the classical description of the motion and thus is best suited for the purpose of arriving at the classical analogue of any quantum theory.

Now, in the theory of the older equation (2), describing particles with the spin of one-half, the Heisenberg equations of motion lead to the idea of the "Zitterbewegung" of the electron. This appears effectively from writing the particle coordinates x_r in the form

$$x_r = y_r + \xi_r, \tag{69}$$

separating it into the part y_r:

$$y_r = b_r + \frac{p_r}{p_0} t, \tag{70}$$

describing the classical motion, the b_r being independent of time t, and the part ξ_r describing the small, high-frequency oscillations. In the present theory there are several possible motions, because here the coordinates x_r can vary in two distinct ways; they can vary with

(i) time
(ii) gauge transformations.

For a particle with zero momentum, the motion under the gauge transformations is found to correspond to the point x_r wandering over a spherical shell. It is of no physical significance. The physically important thing is the spherical shell itself. Its radius vector is found to be $-s_{r0} = +s_{or}$ [see definitions in equations (64)]. The radius vector is found to oscillate with time. Thus we have the picture of a pulsating spherical shell.

What then of the future of the new equation? At the present stage there are some quite serious difficulties, which prevent any further development of this theory. They are connected with the apparent impossibility of having electromagnetic interactions of the particle described by the equations (37) treated by any known methods. The reason for this is that, when one attempts to introduce an electromagnetic field (defined by its four-potentials A_μ) by replacing the four-momenta p_μ in the wave equation by $p_\mu + eA_\mu$,

$$p_\mu \rightarrow p_\mu + eA_\mu, \tag{71}$$

A Positive-Energy Relativistic Wave Equation

one finds that the transformed set (47) of wave equat
not internally consistent. The internal consistency is re
four potentials have the form

$$A_\mu = \partial_\mu S$$

for some function S. That, however, corresponds simp
where all components of the electromagnetic field tensor

$$F_{\mu\nu} \equiv \partial_\mu A_\nu - \partial_\nu A_\mu$$

vanish. Hence no physically observable field exists. It mea
that we have no consistent theory for our particles if they ar
have any other electromagnetic characteristics.

BIBLIOGRAPHY

P. A. M. Dirac. *Proc. Roy. Soc.* (*London*) **A322**, 435 (1971).
P. A. M. Dirac. *Proc. Roy. Soc.* (*London*) **A328**, 1 (1972).

5 — Cosmology and the Gravitational Constant

Introduction by Professor E. P. George

Lecture presented at the School of Physics, University of New South Wales, Kensington, Sydney, Australia, August 27, 1975. Introduction and discussion organized by Professor E. P. George, Head of School.

This will be the third and last of Professor Paul Dirac's lectures in this university. I want to take this opportunity, sir, of expressing our very great thanks to you for coming all this way. In this part of the world we get far too few of the really great physicists, I am afraid. It has been quite an inspiration to us to have you with us here for this last week. I only regret that you could not stay longer!

To say that we have greatly appreciated and have been inspired by your lectures is unnecessary. It is obvious by the enormous interest your lectures have created, by the number of people that turned up at all your lectures.

Not only has the content of your lectures been inspiring but even your method of delivery. I myself and many of my colleagues have learned a few tips, and our lectures, I think, will be much improved by observing the way you organize your blackboard material and the way you speak, without any hesitation. I think that we have all benefited as much from that (or partly from that) as from the actual content of your talks.

Professor and Mrs. Dirac will be leaving shortly after this lecture and there will not be time for any more discussions about physics with anybody here who felt he might have a last-minute question on which he would like some clarification. Professor and Mrs. Dirac leave for Adelaide shortly after this lecture to stay with Sir Mark Oliphant at the Government House there. From there they will go on to the Australian National University in Canberra, again for a week. From there they will go to New Zealand, where Professor Dirac is to give an invited lecture at the University of Canterbury.

Well, once again, sir, many, very many thanks for coming. Please, put a good word in for us when you meet some of your colleagues. We would love to see more of them. So, if you have enjoyed your stay here, as I hope you have, please pass the word around that we would like to see more people like you here!

I shall now call on Professor P. A. M. Dirac, Fellow of the Royal Society, to give his lecture entitled "Cosmology and the Gravitational Constant."

PROFESSOR P. A. M. DIRAC

The argument in favor of the variation of the gravitational constant comes from a study of the constants of Nature. Nature provides us with various constants: the velocity of light, the charge of the electron, the mass of the electron, and quantities like that. Most of these have dimensions, that is to say, the value of the quantity depends on what units you use. When we use the metric system of units, we get a different value for the constant from that obtained when we use the British system. Well, such numerical values are not of any general interest. However, from the constants of Nature, we can construct some that are dimensionless, some that are the same in all systems of units. It is only these dimensionless quantities that we shall be dealing with today.

One of these dimensionless constants is the famous reciprocal of the fine-structure constant

$$\frac{\hbar c}{e^2}. \tag{1}$$

It is fundamental in the atomic theory, and it has the value of about 137. Another dimensionless constant is the ratio of the mass of the proton to the mass of the electron, that is to say,

$$\frac{m_p}{m_e}. \tag{2}$$

That constant has the value somewhere near 1840. At present, there is no satisfactory explanation for these numbers, but physicists believe that, ultimately, an explanation will be found. One would then be able to calculate them from basic mathematical equations. One may expect these numbers to occur as being built up from 4πs and other simple numbers like that.

Now, there is another dimensionless constant of Nature which I want to call your attention to. It arises as follows: Consider the hydrogen atom: electron and proton. The electrostatic force between them is inversely proportional to the square of their distance. So is the gravitational force. We can then take the ratio of the electrostatic force to the gravitational force. It will be independent of the distance, and it will be dimensionless. In this way we get the number

$$\frac{e^2}{G m_e m_p}, \tag{3}$$

where e is the charge of the electron (and proton), G is the gravitational constant, and m_e, m_p are the masses of the electron and the proton.

Now, if we work out its value, we get an extremely large number. It is about

$$2 \times 10^{39}.$$

There ought to be some explanation for this number, like for the other dimensionless constants of Nature. How can we ever hope to be able to get a theory which will lead to such a large number? We cannot build that out of 4πs and other simple numbers provided by mathematics, in any reasonable way! The only way in which one might account for this large number is by connecting it with the age of the Universe.

Now, when I talk about the age of the Universe, I am referring to the generally accepted Big Bang model of the Universe. According to this model, the Universe started out with all the matter concentrated in a very small region, or perhaps even at a mathematical point. There then was a tremendous explosion at the beginning, many chunks of matter being shot out with various speeds. The more rapidly moving chunks would then move farther. That would give the situation which we see at present, with all the matter receding from us and the more distant matter receding more rapidly than the nearer matter. The velocity of recession is proportional to the distance.

Again, one might even adopt the model which was proposed by Lemaître. According to it the Universe started off as a single atom—an atom with a very large mass, all the mass of the Universe. This single large atom was extremely radioactive. It immediately underwent disintegration into various pieces, these pieces underwent further disintegration, the disintegrations continued, and the radioactivity which we see is just the remnants of this initial radioactivity. This is quite a nice picture, but it is not essential for any argument that I am going to give you now.

We have in this way an age for the Universe. This age is called the epoch. One can see what this age is by using Hubble's constant, which connects the velocity of recession of distant objects in the sky with their distance. Hubble found that the velocity of recession is proportional to the distance. He was able to check that accurately only for the nearer objects, but it held so well that it is now assumed to be valid also for the more distant ones. From the ratio of the velocity of recession to the distance, one can determine how long ago in the past all the matter was originally concentrated in a very small volume, and thus one gets the age of the Universe, or the epoch. There are big uncertainties in this estimate of the age of the Universe because of the uncertainties in measuring the distances of very faraway objects. The most recent figure, given by Abell of U.C.L.A., is about

$$t = 18 \times 10^9 \text{ years.}$$

This figure involves years, a rather artificial unit of time. We may use instead a unit of time provided by atomic theory. Let us take as the unit, say, the time required for light to traverse a classical electron:

$$\frac{e^2}{m_e c^3}.$$

If we express t in terms of this unit, we get a number of the order 7×10^{39}:

$$t = 7 \times 10^{39} \frac{e^2}{m_e c^3}.$$

This is a number roughly the same as the large number, 2×10^{39}, which we obtained here previously.

Now, you may say: "This is a very remarkable coincidence". However, I do not believe it is a coincidence; I believe that there must be some fundamental reason in Nature why these two large numbers should be so close together. We do not know that reason at present, we cannot guess at it. It will be explained, however, when we have better information both about atomic theory and about cosmology.

Let us assume, then, that there is a connection between these two numbers, a connection which will be provided by a theory of the future. The number $t = 7 \times 10^{39}$ is not constant; it gets bigger as time goes on. Thus if these two numbers are interconnected, the number $e^2/Gm_e m_p = 2 \times 10^{39}$ must also get bigger as time goes on and must maintain the same ratio to the first number.

We may picture this result most conveniently by adopting a system of units (we may call them "atomic units") such that the electron's charge e, its mass m_e, and the proton mass m_p remain constant when referred to them. Then, since $e^2/Gm_e m_p$ varies with time, G itself must also vary in time, referred to those units. We have

$$\frac{e^2}{Gm_e m_p} :: t, \tag{4}$$

so G must vary inversely proportionally to time,

$$G :: t^{-1}, \tag{5}$$

when referred to atomic units.

I shall be continually referring here to G varying according to t^{-1}. This is always to be understood as meaning that "G, expressed in atomic units, varies according to t^{-1}." G is a thing that has dimensions, and how it varies with time will depend on what unit you use. You must use atomic units in order to have the law $G :: t^{-1}$.

We then have a sort of a general principle that very large numbers which turn up in Nature and have no dimensions, are related to each other. I call this principle the

Large Numbers Hypothesis.

According to it, all the very large dimensionless numbers, which turn up in Nature, are related to one another, just like $t = 7 \times 10^{39}$ and e^2/Gm_em_p.

There is one further very large dimensionless number which we have to take into consideration. That is the total mass of the Universe when expressed in units of, say, the proton mass. That will be, if you like, the total number of protons and neutrons in the Universe. It may be, of course, that the Universe is infinite and that, therefore, this total number is infinite. In that case we should not be able to talk about it. Yet we can use another number to replace it. We need only consider that portion of the Universe which is sufficiently close to us for the velocity of recession to be less than, let us say, half the velocity of light. We are then considering just a certain chunk of this infinite Universe, for which recession velocities are less than half the velocity of light. We then ask, what is the total mass of this chunk of the Universe? That again will be a very large number and will replace the total mass of the Universe, to give us a definite number when the Universe is infinite.

We may try to estimate this total mass using the mass of those stellar objects which we can observe, and making an allowance for unobservable matter. We do not know very well how big that allowance should be: there may be quite a lot of unobservable matter in the form of intergalactic gas or black holes or things like that. Still, it is probable that the amount of dark matter is not very much greater than the amount of visible matter. If you make an assumption of that kind, you find that the total mass, in terms of the proton mass, is

$$\frac{\text{total mass}}{\text{proton mass}} = 10^{78}, \tag{6}$$

with a suitable factor allowed for the invisible matter. We, therefore get a number which is, roughly, the square of t (in atomic units).

Now, according to the Large Number Hypothesis, all these very large dimensionless numbers should be connected together. We should then expect that

$$\frac{\text{total mass}}{\text{proton mass}} = 10^{78} :: t^2. \qquad (7)$$

Using the same argument again, we are therefore led to think that the total number of protons in the Universe is increasing proportionally to t^2. Thus, there must be creation of matter in the Universe, a continuous creation of matter.

There have been quite a number of cosmological theories working with continuous creation of matter. A theory like that was very much developed by Hoyle and others. The continuous creation which I am proposing here is entirely different from that. Their continuous creation theory was introduced as a rival to the Big Bang theory, and it is not in favor at the present time.

The continuous creation which I have here is essentially different from Hoyle's continuous creation, because Hoyle was proposing a steady state of the Universe, with continuous creation to make up for the matter which is moving beyond our region of vision by the ex-pansion. In his steady-state theory, he had G constant. Now, in the present theory, G is varying with time, and that makes an essential difference.

I propose a theory where there is continuous creation of matter, together with this variation of G. Both the assumption of continuous creation and the variation of G follow from the Large Numbers Hypo-thesis.

This continuous creation of matter must be looked upon as something quite independent of known physical processes. According to the or-dinary physical processes, which we study in the laboratory, matter is conserved. Here we have direct nonconservation of matter. It is, if you like, a new kind of radioactive process for which there is noncon-servation of matter and by which particles are created where they did not previously exist. The effect is very small, because the number of particles created will be appreciable only when we wait for a very long time interval compared with the age of the Universe.

If there is new matter continually being created, the question arises: "where is it created?" There are two reasonable assumptions which one might make. One is that the new matter is continually created throughout the whole of space, and in that case, it is mostly created in intergalactic space. I call this the assumption of

additive creation.

Alternatively, one might make the assumption that new matter is created close by where matter already exists. That newly created matter is of the same atomic nature as the matter already existing there. This would mean that all atoms are just multiplying up. I call that the assumption of

multiplicative creation.

There are these two possibilities for the creation of new matter. I do not know which to prefer. One should continue with both possibilities and examine their consequences.

Now, if we want to develop this theory with G varying, we have to face the problem of how the equations of mechanics are to be altered. We have Einstein's theory governing gravitation. That theory is extremely good; it has many successes, and we want to preserve those successes. Yet, according to the Einstein theory, G has to be constant: there is no room in Einstein's theory for variation of G. In fact, if you use natural units, you would have G equal to 1.

How can we then allow G to vary and yet retain the successes of the Einstein theory? It seems to me there is only one way of doing that. Namely, we suppose that the equations of the Einstein theory are good equations, correct equations, but that they apply to quantities referred to units which are different from atomic units.

We have to consider distances and times which, in the Einstein theory, are represented by

$$ds,$$

the interval between two neighboring points. We have to consider two expressions for ds: one ds which is valid for the Einstein equations, another ds which would be measured by atomic units. Let us call the Einstein ds

$$ds_E$$

and the ds measured in atomic units

$$ds_A.$$

We assume that with both systems of units we have the velocity of light equal to 1, so that an interval of time or an interval of distance is changed in the same ratio when we go from ds_E to ds_A.

A good many people have been working in recent times with gravitational theories in which G varies, but they all tend to use a primitive theory, according to which inertial mass and gravitational mass are different. G is then the coefficient connecting the two masses, and you can then easily have G varying by having the ratio of gravitational mass to inertial mass varying. Such a theory is really a very primitive theory and it is most unsatisfactory because it goes completely against the Einstein theory. According to the Einstein theory, gravitational mass and inertial mass *have* to be the same. If you adopt the primitive theory, you lose the successes of the Einstein theory; you lose the possibility of explaining the motion of the perihelion of Mercury. I refer you to this primitive theory just because, if you are reading the literature, you will find papers written where calculations have been made with it. I find such calculations quite unacceptable, because they involve abandoning the successes of the Einstein theory.

I use, instead, this theory of two measurements for ds. One may call this

Milne's hypothesis.

Milne was the first to introduce the idea that there might be two scales of time which are important in physics and to discuss the connection between them. He wrote books on the subject before the war. Milne's work is quite independent of the kind of argument I have been giving here about the large numbers. He had various philosophical arguments, which I do not find very convincing. He did have, however, the basic idea that there could be two measurements of time which are important to physics, and that is the idea which I am going to introduce now, in order to be able to maintain the successes of the Einstein theory in spite of the variation of G.

Now, with these two dss (ds_E and ds_A), we shall work out first of all the relationship between them. Let us take a simple example, say of the Earth moving around the Sun in a circular orbit (as an approximation). Then, on using the Newtonian theory, we have the equation

$$GM = v^2 r,$$

where M is the mass of the Sun and v and r are the orbital speed and orbital radius of the Earth. This elementary Newtonian theory is good enough for the present calculation. This equation must be valid whatever units one may use.

I will use the terminology

$$ds_E \rightarrow \text{"mechanical" units}$$

or units which we have to take into account to have the Einstein equations valid. ds_A, on the other hand will be in

$$ds_A \rightarrow \text{"atomic" units,}$$

units provided by atomic clocks or the spacings of crystal lattices.

The equation $GM = v^2 r$ must be valid in both mechanical units and atomic units. Now in mechanical units we have the ordinary idea of mechanics. Every one of G, M, v, and r is a constant. The Earth moves steadily in its orbit at a constant velocity and with a constant orbital radius.

How is it in atomic units? In atomic units we have

$$G :: t^{-1}.$$

With additive creation, the mass M of the Sun (measured in atomic units, say in terms of the proton mass), will be a constant:

$$M :: 1 \text{ (additive creation).}$$

With multiplicative creation, again in atomic units, we have

$$M :: t^2 \text{ (multiplicative creation),}$$

because then every bit of matter is multiplying up according to the t^2 law.

Thus the way in which M varies with time, in atomic units, depends on the assumption we make about the mode of the creation of new matter.

What about v? Well, v is dimensionless; we can express it as a certain fraction of the velocity of light. It must be the same whether we use mechanical units or atomic units: v is the same fraction of the velocity of light in both cases. Thus v does not vary:

$$v :: 1.$$

Then, on comparing the two sides of the equation: $GM = v^2 r$, we see that

$$r :: t^{-1} \quad \text{(additive creation)}$$

and
$$r :: t \quad \text{(multiplicative creation)}.$$

Thus, referring to atomic units, we see that the radius of the Earth's orbit is getting less, if we have additive creation—the Earth is approaching the Sun. The same argument applies generally to all the distances in the Solar System. Our Solar System is to be imagined as contracting, on that picture. This must be a cosmological effect, superposed on any other effects, arising from known physical causes. Similarly, if we adopt the assumption of multiplicative creation, the Earth is moving away from the Sun, and all the distances in the Solar System are expanding. This is, again, a cosmological effect, independent of any known physical processes.

Well, there we have effects which we might hope to be able to measure, and so check up on whether this theory is a good theory or not. We just have to make accurate observations with atomic time. I should emphasize that it is important that these observations are made with atomic time, because the above formulas apply only to quantities measured in atomic units.

We might, first of all, think of the Moon and make observations of the Moon to check on this theory. Now, people have been making observations of the motion of the Moon for the last 20 years with atomic time. They have also recently been making accurate observations of the distance of the Moon, referred to atomic units. The astronauts who landed on the Moon put down some laser reflectors, and people are now sending laser light to these reflectors and observing the light reflected by them. They then measure, using an atomic clock, the time taken by the light to get to the Moon and back and, in that way, get the distance of the Moon, referred to atomic units.

If we apply it to the motion of the Moon around the Earth, our theory would require that with additive creation the Moon should be approaching the Earth by an amount we can easily calculate. It is about

$$2 \text{ cm/year}.$$

With multiplicative creation, the Moon should be moving away from the Earth at the same rate. We would have to measure, therefore, the distance of the Moon to that accuracy. Now, people have recently been measuring the distance of the Moon with very great accuracy. The most recent information I obtained was that, nearly a year ago, they had the

Moon's distance to an accuracy of 6 cm and that they were continually improving on that. It is just a question of waiting a little while and making further observations.

There one would think that we have a definite method of checking this theory. Matters are not, however, as simple as that. The Moon's motion is very strongly influenced by the tides. Now, tidal effects are large compared with these effects which we are measuring. They cannot be calculated with sufficient accuracy, so far as I know. I had a talk last year with Jim Williams of the Jet Propulsion Laboratory. He was rather pessimistic about the possibility of calculating the tidal effects with sufficient accuracy so as to be able to check this theory. I do not know how far people will be able to progress with that. It is something which is not hopeless. I hope to be able to have a talk with people here who are doing the lunar ranging and to see what they say about it.

We have there one method of checking this theory. We might take another method involving not the distance of the Moon but the angular motion of the Moon through the sky. Let us use the symbol

$$n$$

for the angular velocity of the Moon in the sky. Then we can get the (relative) angular acceleration:

$$\frac{\dot{n}}{n}.$$

People can make very accurate observations of this angular motion of the Moon in the sky by observing the times of lunar occultations of the stars. For the last 20 years people have been observing these occultations with atomic clocks. First, the observations were visual; the more recent ones are entirely automatic. Thus there is a possibility of working out $\dot{n}/n$ from these observations of the motion of the Moon.

Several people have been working on this question. The man who has done most on it, so far as I know, is Van Flandern, in the Naval Research Observatory in Washington.

The problem is to observe this lunar acceleration, referred to atomic time:

$$\left(\frac{\dot{n}}{n}\right)_{\text{atomic}}$$

and to observe it also, referred to the standard time used by astronomers, which is called the ephemeris time. This is the time marked out by the motion of the Earth around the Sun or the motion of the planets. It need not be the same as atomic time. Its units are set up just by the Newtonian equations of motion, or the Einstein equations, if one wants to proceed to greater accuracy.

We can then take the difference of $\dot{n}/n$ referred to these two time units. We have

$$\left(\frac{\dot{n}}{n}\right)_{\text{atomic}} - \left(\frac{\dot{n}}{n}\right)_{\text{ephemeris}} = \left(\frac{\dot{n}}{n}\right)_{\text{difference}}.$$

Van Flandern made observations of this quantity, which gave

$$\left(\frac{\dot{n}}{n}\right)_{\text{difference}} = (-16 \pm 10) \times 10^{-11}/\text{year}.$$

Now, Van Flandern did his early calculations with respect to the primitive theory of gravitation which I was talking to you about, the theory which involves treating gravitational mass and inertial mass as two independent things. According to that primitive theory,

$$\frac{\dot{G}}{G} = \frac{1}{2}\left(\frac{\dot{n}}{n}\right)_{\text{difference}}. \tag{8}$$

On Van Flandern's data this would give

$$\frac{\dot{G}}{G} = (-8 \pm 5) \times 10^{-11}/\text{year}. \tag{9}$$

Van Flandern was rather pleased with that result, because the theory with $G :: t^{-1}$ gives

$$\frac{\dot{G}}{G} = -\frac{1}{t},$$

which is then equal to about

$$-6 \times 10^{-11}/\text{year},$$

corresponding to the reciprocal of our latest estimate of the epoch:

$$t = 18 \times 10^9 \text{ years}.$$

However, this is not very satisfactory from my point of view because it involves the primitive theory.

If we replace it by the theory based on Milne's Hypothesis, we get, with additive creation,

$$\frac{\dot{G}}{G} = -\left(\frac{\dot{n}}{n}\right)_{\text{difference}} \quad \text{(additive creation)}$$

and, with multiplicative creation, we have

$$\frac{\dot{G}}{G} = +\left(\frac{\dot{n}}{n}\right)_{\text{difference}} \quad \text{(multiplicative creation)}.$$

In connection with Van Flandern's data, these then give

$$\frac{\dot{G}}{G} = (16 \pm 10) \times 10^{-11}/\text{year (additive creation)}$$

and

$$\frac{\dot{G}}{G} = (-16 \pm 10) \times 10^{-11}/\text{year (multiplicative creation)}.$$

Well, we want $\dot{G}/G$ to be negative, as follows at once from the formula $\dot{G}/G = -1/t$. So we see that Van Flandern's observations support the idea of multiplicative creation. They give rather too large an effect: -16 instead of -6.

Van Flandern has been continually checking and rechecking his calculations and has been modifying his results somewhat. His most recent results that I have heard about are considerably less than his original 8 [in $(-8 \pm 5) \times 10^{-11}/\text{year}$], and they are getting closer to what the theory wants.

I should perhaps explain the immense complexity of these calculations. You have to make very accurate calculations of the motion of the Moon, and the Moon's motion is very complicated. The Moon is just one of the members of the Solar System and is being continually perturbed by all the other planets. You might be inclined to think that other planets, being far away, would not have much effect. When people look, however, into things more closely, they find that an effect which they were previously assuming to be negligible is not at all negligible. It has to be taken into account, and it does affect the results. People are continually working on the effects of the other planets and, as the result of this, Van Flandern is changing his figure.

I think one must wait a little longer and see what final result comes from this work. In any case, you see that it is within the reach of present-day technology to check on the validity of this theory. One cannot yet say that the results really confirm the theory: it is a bit too soon to say that is the case. One can hope, however, to confirm or disprove the theory definitely within a few years.

We do have, of course, tidal effects coming in also here, but in the case of these calculations of the relative angular acceleration of the Moon, they cancel out as they apply both to atomic and ephemeris observations.

One can also make observations using other planets, not the Moon. Now, if you get away from the Moon, you get away from the disturbance produced by tidal effects. One can make very accurate observations of the distances of other planets by radar. I. I. Shapiro is working on this. The method is to send radar waves to a planet and to observe the reflected waves. These reflected waves are, of course, extremely weak, but there is a big transmitter and reflector of radar at Arecibo in Puerto Rico which is used for this work, and it is sufficiently sensitive. This radar receiver at Arecibo has recently been refitted (six million dollars it cost, I believe) to improve its accuracy very strongly. During the last few months it has been used to observe Venus. I have not heard what results they have obtained. Venus is just approaching very close to the Earth at present, and they hope to make sufficiently accurate observations of Venus to show up even dried riverbeds, if there are any there. These observations will enable us to obtain the distance of Venus with very great accuracy. If, then, one continues the observations for a few years, one will be able to determine whether these distances are varying, as they should according to this theory. Similar observations could be made with other planets.

I think Shapiro's observations are, perhaps, more hopeful than Van Flandern's, because in dealing with the planets we do escape the difficulties brought on by the tides. Shapiro has been working on this for some years, and he is rather hesitant about giving results before they are very well established. I have heard indirectly that his observations tend to support additive creation and that they would be, therefore, in conflict with Van Flandern's. However, I think that we must wait a year or two for Shapiro to get new observations with this improved apparatus at Arecibo. Then we shall have another check of this theory.

Thus there are at present three astronomical ways in which one may check this theory of the variation of the gravitational constant. They concern the distance of the Moon, the motion of the Moon, and the distances of the planets. I think that in a few years we shall get a definite answer.

If this theory is verified and accepted, it will mean quite a big alteration in our ideas of cosmology. Cosmologists tend nowadays to favor a picture of the Universe where it is expanding at the present time, but the expansion is slowing down and will ultimately stop and change into a contraction.

That kind of picture of the Universe would not be permissible according to the ideas I have been presenting here. It would not be permissible because it would lead to a constant large number, namely, the maximum size of the Universe, a large number which does not vary with the epoch. The maximum size of the Universe, of course, has nothing to do with the epoch. Thus we should get a large number which contradicts the Large Numbers Hypothesis.

According to my views, there can be no maximum size to the Universe: it must go on expanding, forever. And G will correspondingly go on getting weaker and weaker, forever.

Now I will stop at this point and may be able to have time to answer one or two simple questions.

PROFESSOR E. P. GEORGE

I invite you now to address your questions to Professor Dirac.

Question (from Professor J. C. Kelly)

Sir, in your equation: total mass/proton mass $= 10^{78}$, you have chosen to keep the proton mass constant and vary the total mass of the Universe. Why do you not allow the mass of the proton to change?

Professor Dirac

You can, but it does not mean anything saying that the mass of the proton changes. You must only take a dimensionless number and ask whether that changes or not. When I talk about the mass of the Universe changing, that is to be understood as the mass of the Universe referred to atomic units. Just as when I talk about G changing, it means G referred to atomic units. If anything is not dimensionless then you cannot discuss whether it changes or not.

Question

Is the atomic time a time which involves Planck's constant?

Professor Dirac

Yes. Atomic time would be the time measured, for example, by the caesium clock or some other atomic clock.

Question (same questioner)

Yes? Is that not rather different from the unit you wrote down at the beginning of the lecture?

Professor Dirac

e^2/m_ec^3? No! It would be connected with it by a 137 or some such factor. Not really independent. No!

Question

Sir, as G is getting smaller, does the ratio of the strength of gravitational forces to that of electrical forces get smaller?

Professor Dirac

It gets smaller! Yes!

Question (the same questioner)

So the gravitational forces get weaker?

Professor Dirac

Yes! That is how one is to understand their being so weak at the present time. They have had such a long time in which to get weaker!

Question

Sir, you discussed at length large numbers. Is there any evidence to support the hypothesis that any of the smaller numbers, such as 137 or so on, might also be changing with time?

Professor Dirac

It is possible that they might change logarithmically with the time. There is observational evidence that the number 137 does not change logarithmically with the time. There is no evidence about the number $m_p/m_e \simeq 1840$. It would be very interesting to know whether this number is really constant or whether it varies logarithmically with the time. This is something that could be measured with the present-day technology. The way you would measure it would be to compare two atomic clocks in one of which the time is marked out by the vibration of an electron and, in the other, the time is marked out by the vibration of some atom. For instance, you might have a caesium clock where the time is marked out by oscillations of electrons and an ammonium (or methane) clock, where the time is marked out by the oscillations of some atomic nucleus.

Now, I was told that these atomic clocks enable one to measure time to within an accuracy of one part in 10^{13}.

If there is a logarithmic variation with time in the quantity m_p/m_e, there will be a logarithmic variation in the ratio of the times, as measured by the two kinds of clocks. This would be a hundred times less than the variation in G and would be of the order of one part in 10^{12} per year.

Now, if you can measure to the accuracy of one part in 10^{13}, you should be able to detect a difference of one part in 10^{12}. So it is just a question of developing the technology so as to be able to compare these two clocks with the necessary accuracy, and then actually making the comparison for a year or so.

I hope people will do this experiment, because it seems to me to be very important to decide whether the number m_p/m_e is really constant or not. It is such a mysterious number, and no one has any explanation for it. That would be a beginning.

Question

Just a comment for the ABC News; in April of this year, I think, there was a considerable doubt whether the Arecibo reflector would be of suitable quality to get that increased sensitivity for the Venus conjunction in the latter part of this year.

Professor Dirac

I was assured, even more recently, that it is successful and that it is giving very good results.

Question

You stressed throughout that those effects lie outside the normal physical laws. Then, is there any assumption that this new material that appears is quantized or could it be a different form of matter?

Professor Dirac

It must be quantized, because we don't notice any other kinds of matter present. It must be indistinguishable from ordinary matter.

Question

The calculation of the age of the Universe was based on the assumption of the Big Bang theory, I presume, whereas we also have built into this theory a continuous creation. I am wondering whether the Big Bang is, in fact, necessary. It seems to me that one can end in conflict when one has the original Big Bang, together with the continuous creation, whether it be additive or multiplicative.

Professor Dirac

I don't see any conflict there. In any case, you have the Big Bang because you have the various spiral nebulae receding from one another with velocities proportional to their distance. If you go back in time, then, at a sufficiently remote time in the past, they would all have been very close together. That would have been the time when the Universe started.

Question (from Professor P. Fisher, Wollongong University)

In these processes one requires nonconservation of matter. Doesn't this also imply nonconservation of linear momentum . . . ?

Professor Dirac

That is so! But these questions, of whether you have conservation of something which is not dimensionless, don't have very much meaning. If a quantity has dimensions, then, whether it varies or not depends on what units you use. Thus whether it is conserved or not depends on what unit you use. You could say that: "referred to mechanical units, these things are conserved—referred to atomic units, they are not conserved."

Question

Would an Eötvös-type of an experiment have any bearing on your conclusions, such as was done by Dicke in recent years?

Professor Dirac

Yes! It would be possible, perhaps, to measure G by terrestrial experiments. It would be necessary, however, to have an enormous improvement in accuracy. I think they can measure G only to an accuracy of one part in 10^6 at the present time. There are prospects for making vast improvements in that.

Question (from Dr. H. Murdoch, Sydney University)

Is it worthwhile to extrapolate back towards $t = 0$? If G is the only quantity which varies with time, as far as the forces go, what could possibly cause the Big Bang, since, as is easily shown, G would extrapolate back to a figure approaching infinity?

Professor Dirac

You can't go too far back into the past because other physical quantities may be varying. m_p/m_e, for example, may be varying. The constants which come into nuclear interaction may vary. And we don't really know anything about that. It becomes very uncertain when you try to go back far into the past.

Question (from Dr. H. Murdoch, Sydney University)

This seems to indicate that a lot of things must be varying or else the Big Bang would not have occurred.

Professor Dirac

May be a lot of things are varying but, in any case, we don't have any theory of how the Universe started. It is just speculation.

Professor E. P. George

There are obviously going to be many, many more questions, coming from the audience than, I think, Professor Dirac will have time to answer. I ought to take, perhaps, two questions from this side of the room and two from that side of the room, and then I am afraid, we probably ought to let Professor Dirac get away to Adelaide.

Question

I was just going to ask whether the attempts by Eddington to explain the numbers, the cosmological constants, still have got any validity.

Professor Dirac

I don't think so! He had some arguments to provide for the number 137. Yet this 137 is still a complete mystery. He also had arguments to provide for the total number of particles in the Universe: 2^{256}, I think. Well, that would be in direct conflict with the present theory, because, according to the present theory, that number should not be a constant at all, so it just can't be 2^{256}!

Question (from Dr. H. Murdoch, Sydney University)

Your theory requires the total number of particles in the Universe to increase with time. Now, in the standard cosmological theories, with the expansion slowing up, there is an increase with time in the total number of particles observable. This comes from the McVittie treatment, for example, of horizon theories. Does this increase have a significant effect on the amount of matter required in your theory?

Professor Dirac

I don't think it could be very significant, because my theory would not be consistent with the slowing up of the expansion.

Question

I was wondering if a suitable combination of additive creation and multiplicative creation could give any answer you'd like?

Professor Dirac

That is so! But it would be rather unreasonable, I think. I don't suppose anybody would believe in a combination of both additive and multiplicative creation.

Question

What is the souce of your created matter: is it an excess energy or something else?

Professor Dirac

No! It is to be thought of as a sort of a radioactivity. These things are created out of the vacuum by some new kind of radioactive process, quite independent of any that are already known and which occurs too seldom to show up in ordinary laboratory experiments.

Professor E. P. George

I am afraid that has to be it! If we don't let Professor Dirac go now, we will get a stiff note from the Governor of South Australia!

Index